The Concept (
Modern Thec

This book examines the importance of the animal in modern art theory, using classic texts of modern aesthetics and texts written by modern artists to explore the influence of the human-animal relationship on nineteenth and twentieth century artists and art theorists. The book is unique due to its focus on the concept of the animal, rather than on images of animals, and it aims towards a theoretical account of the connections between the notions of art and animality in the modern age. Roni Grén's book spans various disciplines, such as art theory, art history, animal studies, modernism, post-modernism, posthumanism, philosophy, and aesthetics.

Ph.D. Roni Grén is a researcher at the University of Turku, Finland. His topics of research have been centered on modern art theory and French culture. Formerly, he has written a book on Georges Bataille's writings on art, and is currently working on research entitled "Prehistory and Modernism."

Routledge Advances in Art and Visual Studies

A full list of titles in the series can be seen at: www.routledge.com/Routledge-Advances-in-Art-and-Visual-Studies/book-series/RAVS

Intersecting Art and Technology in Practice
Techne/Technique/Technology
Edited by Camille C Baker and Kate Sicchio

Wonder in Contemporary Artistic Practice
Edited by Christian Mieves and Irene Brown

W.J.T. Mitchell's Image Theory
Living Pictures
Edited by Krešimir Purgar

The Politics of Contemporary Art Biennials
Spectacles of Critique, Theory and Art
Panos Kompatsiaris

Contemporary Visual Culture and the Sublime
Edited by Temenuga Trifonova

Art, Animals, and Experience
Relationships to Canines and the Natural World
Elizabeth Sutton

What Drawing and Painting Really Mean
The Phenomenology of Image and Gesture
Paul Crowther

The Concept of the Animal and Modern Theories of Art
Roni Grén

The Aesthetics of Scientific Data Representation
More than Pretty Pictures
Edited by Rikke Schmidt Kjærgaard and Lotte Philipsen

The Concept of the Animal and Modern Theories of Art

Roni Grén

Routledge
Taylor & Francis Group
LONDON AND NEW YORK

First published 2018 by Routledge

2 Park Square, Milton Park, Abingdon, Oxon OX14 4RN
605 Third Avenue, New York, NY 10017

Routledge is an imprint of the Taylor & Francis Group, an informa business

First issued in paperback 2022

Publisher's Note

The publisher has gone to great lengths to ensure the quality of this reprint but points out that some imperfections in the original copies may be apparent.

Library of Congress Cataloging-in-Publication Data
A catalog record for this title has been requested

ISBN: 978-1-138-05426-4 (hbk)
ISBN: 978-1-03-233953-5 (pbk)
DOI: 10.4324/9781315163628

Typeset in Sabon
by Apex CoVantage, LLC

For Tati

Contents

Figures

Acknowledgments

This work was started about five years ago. The process has been quite long, mostly rewarding, but from time to time troublesome. I would first of all like to thank Niina for her invaluable presence in my life. To Professor Altti Kuusamo I am indebted for much, for his comprehension and the unconditional support my research work has received from him the whole decade. The same can be said of the whole staff of the Department of Art History of the University of Turku. For the editorial assistance I've had, I would especially like to thank Julia Michaelis and Isabella Vitti from Routledge. For the excellent remarks made on my manuscript, my gratitude belongs to Chris Turner-Neal. My work has been funded by Kone Foundation and the School of History, Culture, and Arts Studies of the University of Turku.

Turku, December 23, 2016
Roni Grén

Acknowledgments

Introduction

The subject of this book might at first sight seem to be marginal, if not superficial. Artistic modernity is not known for its animal imagery, nor for its empathic attitude towards nature. On the contrary, modernism has been repeatedly accused of reluctance to deal with natural facts and disinterest in ecological questions. These accusations are completely justified, but, as I also hope to show, modernist theory has found different ways to provide space for the animal. Often, its conception of the animal was even used, either positively or negatively, to determine what was art and what was not, in the sense given by the context. These theories mostly suggested a fixed way of defining animality in relation to human activity. This is why the work at hand concentrates on the conception of animality and its evolution: the boundaries with which animality was defined were essentially interconnected to central questions of modernity that at first may have seemed to be quite distant from animality.

In this regard, my work aims to answer the following questions:

(1) *What was the status of the animal in the modern theory of art?*
(2) *What was the animal's task in modernist art, and how was the task related to the modernist notion of art?*
(3) *If the animal had a task, by what means was the image animalized; how was the animal assimilated into being a part of the prevailing conceptions of a modernist image?*

From the start, a few caveats must be stated. Although I am mostly critical towards the positions taken by the theories I introduce, my aim also is to show how the concept of the animal developed and how as such it was evaluated against the background given by the natural scientists of the time. This does not mean that I intend to report on the progressivist tendencies of modernity, a report finally culminating in the posthumanist philosophies of our time; quite the contrary, I would prefer to show how several forms of resistance against the knowledge acquired by natural sciences determined the readings, and despite the fact that the forms of this resistance vary historically, they are still continuously in play. The concept of the animal is always, more or less, a consequence of this resistance.[1]

Another point that should be made touches upon situating the work against the increasing amount of literature handling the topic of contemporary animal art. Most of the material connecting itself with this current is openly ideological in approving the contemporary works and the theories behind them, while giving a negative value to most of the efforts made in the context of high modernism. The most notable of these recent contributions within art historical literature has been Steve Baker's work *The Postmodern Animal* (2000). Baker argues quite strongly in the book that the "animal is the very first thing to be ruled out of modernism's bounds."[2] Baker's idea is that modern art was forced to abandon "the animality" of the animal, even in those pictures that showed animals. His point of view is based on the fact that the modern tradition was founded on aesthetic ideals that didn't give as much importance to the represented subject, giving its preferences either to the process of artistic expression (the legacy of Romanticism) or to the autonomical structure of the surface (the tradition of modern formalism, and of the linguistic turn). Thus, the representation of the animal itself, if represented at all, could be still considered secondary.

Without assuming a political pose against this current—rather, I salute with great pleasure most of the recent developments, as well as their ability to support discussions on ecological subjects and animal rights in the domains reserved for exhibiting art—I must still make a critical note against its most severe judgments concerning modernity, through which it undervalues the important role that the concept of the animal, in one way or the other, has played in art theoretical literature. This does not mean that I would reduce postmodern criticism to a mere platitude. If and when the modern attitude is to be blamed and criticized, the new attitudes should also be able to recognize their own historicity. This can only be done by understanding that the criticism itself can merely build on some foundation in which the animal already plays a certain role. My point is that the modernist theory had in many ways connected its image of the animal to notions such as creativity, otherness, matter, or artistic gaze, even as it defined human subjectivity and consciousness with regard to its notion of the animal. The modernist poses made in this regard enjoy a great variety.

As far as I know, my work has no direct precursors. Of course, there are other works that attempt to clarify the meaning of the animal in the modern age in regard to art, but they often tend to discuss the animal question in a different light—for example, in the light of the history of animal rights.[3] Also, there do exist some notable attempts to deduce from the history of the modern theory of art a certain tendency to see the world from an animal perspective—Margot Norris's great work *Beasts of the Modern Imagination* (1985) should be mentioned for its tracking of the "biocentric tradition." Similarly, there are fine accounts of at least some of the important theoretical tendencies that changed the course of the modern appropriation of the concept of the animal to the artistic field: works on Darwinism, fine research articles on particular art movements, especially concerning Surrealism, and of course a multitude of texts pointing out some aspects of the topic with regard to individual artists, but there still is a lack of a larger theoretical account of the history of modernity.

A reason for such a lack, at least a partial reason, might be one that also haunts the arguments of this book. The few theories of the animal, written in the context supported by modern aesthetics, are in a highly disproportionate relation to the multitude of animal images produced during the era. This is not only a consequence of the fact that animal art has long been considered as being of secondary importance and that it often was connected with the sort of sentimentality the modern avant-garde discriminated against (in this sense, of course, avant-gardistic modernity, by rejecting the academic hierarchies, meant a potential ascension for animal art within the cultural hierarchy), but also because the theoretical views on animal life were often established in relation to a context in which only humanity, and its assumed perspective really mattered. Thus, animality was brought up only because of the otherness that it could be made to represent, and as such the theoretical habit of referring to the animal often demanded no counterpart in visual practices. The birth of the philosophical theory of art, the birth of aesthetics, was of major importance in the development. Whatever the reasons (we shall consult some of them later on), the fact that the theories made a problem of the concept of the animal—a concept in the sense in which Jacques Derrida has spoken of "*animot,*"[4] a word denoting all of animality, from bacteria to mammoths, a term signifying a certain form of otherness, and a concept that lives its own cultural life regardless of the differences understood to mark the boundaries between the species—can begin to explain the difficulties in examining the topic, since the concept is hard to illustrate with regard to individual works of art. Properly speaking, such use of the concept of the animal also makes the whole field under discussion quite estranged from contemporary theories about the animal and art, which mainly tries to discuss animals, plural, under the name of "nonhuman animals"—such a term is not only anachronistic to the modernist field but misses the point at which it became necessary for many of the theorists to determine animality as simply that which is not human and has the nature of the other.

Due to some of the difficulties in question, my work deals with several kinds of textual material, some theoretically less plausible than others in light of our contemporary knowledge. The diversity of the material has already given birth to some criticism of my work, so I shall at least try to explain the reasons for my selections. Since one of my principal goals has been to show how the concept of the animal was conceived in the context of modern art, where it was brought in touch with philosophical theories about art, my work requires an interdisciplinary point of view. This has necessitated inclusion of heterogeneous material in the same work: starting with the historically important metaphysical views touching upon the animal, I end by tackling the most vulgar interpretations of Darwinist naturalism written in the context of avant-garde art. Conveniently, I have not tried to make self-explanatory remarks on the reasons for integrating such and such arguments into my work, but I have tried to give the reader an understanding of their interrelatedness for the sake of giving a picture of those aspects I thought most important to the appropriation of the concept of the animal in the theoretical discourse on the arts, in the shape that we understand as being fully modern.

Additionally, during the course of my research, I was, little by little, confronted by a need to enlarge the frame in which I would discuss the modern theories of art, as I call them. My somewhat large understanding of the concept of the modern is thus partly a consequence of the answers I was starting to find while writing the work. My understanding of artistic modernity extends to mid-18th century France—in this sense, my work pays homage to such writers as Michael Fried and Norman Bryson, who have given art historians permission to think of the context in question in terms usually connected with later periods of modern art.[5] In some ways, this is also a compromise: while the birth of artistic modernity is often seen as occurring in the latter half of the 19th century, the philosophical discussions on the modern conception of the animal still carry within them arguments for and against René Descartes. As the speculative theories on art were earning their place inside the academic disciplines only in the mid-18th century, as meanwhile in France the discipline of art criticism developed, I have no reason to extend my own discussions, aside from small concessions, far beyond it. I have decided to start the work by giving an account of the forms of resistance against the Cartesian paradigm, a resistance that also marked the 18th- and 19th-century views of the human practice of art.

My definition of the modern ends in the frame introduced in the 1960s and the 1970s. This concerns mainly the conception of the animal. Despite my insistence that there exists in this sense a continuity between the modern and the postmodern—an evident fact betrayed by the nature of the terms themselves—I confine my work to the discussions held before the upsurge of the postmodern (and posthumanist) discourse. The reason for this is, predominantly, that the discourse in question has been thoroughly analyzed by the experts in the field, and in the context of my research, I have thought it better to show only the most necessary continuities between the two. This has also given me the chance to stick more clearly to the analysis of the modern discourse by keeping a distance from later developments. The continuities and differences, nevertheless, are briefly discussed in the conclusion of my work.

Some motivation should also be given, at least to warn the reader of what is ahead, of the order of progression. In Part 1 of the work I introduce "the thesis of human exception"—a term introduced by Jean-Marie Schaeffer in relation to the modern philosophical tradition.[6] As a whole, Part 1 goes through several of the major philosophical theories that have affected the modern view on the arts by trying to explain the status of the concept of the animal in them. Part 2 of the work introduces, somewhat anachronistically for my purposes, mid-18th century French discussions, leading the reader from philosophical ideas presented by Étienne Bonnot de Condillac and Jean-Jacques Rousseau to essays by Denis Diderot. Part 2 introduces the question of the origin of the connection between animality and visual art, which makes up one of the main arguments of my work.

Part 3, in its turn, is also quite different from the previous two by its nature and structure. It attempts to list some of the most important characteristics of the questions concerning animality in the 19th century with regard to art theory. The text follows the guidelines given by Darwinism and its appropriation in the visual arts. Darwinism evidently was the most important influence among the changes concerning the understanding of the concept of the animal during the century, but I have also tried to open the scope to complementary developments. Part 3 ends with a somewhat long subchapter discussing the nature and the status of the concept of the animal in Nietzsche's philosophy, which serves as an introduction to the avant-garde theories introduced in the fourth part of the work. Finally, Part 4 of the work leaves the philosophers and the natural scientists almost completely behind and shows how the avant-garde theorists of art integrated philosophical ideas on animality into their systems and manifestoes.

This means broadly that my work proceeds from the canonized philosophical formulations, works and systems canonized by the historiography of philosophy, of the concept of the animal and its possible connections with the essence of the arts, to the adaptations made of them in the artistic field. The philosophical conception of the animal was integrated into the discourse on the arts only when philosophical speculation on art's nature, characteristic to the modern discourse, started to gain ground, i.e. in the mid-18th century. This has forced me to give philosophical views on the animal as background for my arguments.

Nevertheless, I do not argue that the influence was only unidirectional. Although it must be admitted that the hierarchical relations between the purely philosophical and the other discourses on the arts—whatever they might be—have been highlighted in the modern era, there is no need to suggest that the influences could not have gone the other way around, at least in some of the cases introduced. Moreover, the adaptations made by the theorists of art, the ones acting near to the artistic avant-garde, were often such that they altered the original implications of the philosophical theories they interpreted. Moreover, this must have affected the general reception of the theories, at least in the artistic field. More importantly, it has in many cases played a role in the interpretations made of the original theories when they were criticized by later generations.

My own background as a researcher is in the work I have produced in connection with the Surrealist theories on art, mostly concentrating on the works of the now-and-then Surrealist philosopher Georges Bataille and the artist Man Ray. One of the reasons for my writing this work, and for my interest in the topic at all, was my translation of Bataille's posthumously published work *Théorie de la religion* into Finnish. In his work, as I shall explain at more length in Part 4 of the work, Bataille gives an account of the universal history of religion as he conceives it. One of the main arguments of his book is that the history of religion has as its ultimate and as its original form a human need to get away from the human condition by means of a "poetic lie of animality."[7] Bataille was perhaps formulating his concept with a wink, but the theory also gave me a reason to examine more closely whether the animal had had an important role in at least some of the tendencies of the modern avant-garde and its theories. It is true that this role has been ambiguous and has from time to time even been left unnoticed. However, what Bataille also had in mind when writing his theory was that when animality has been brought to the stage, it has carried within itself the ability to open our thinking to what is possible to conceive in terms we use to think of ourselves.

Notes

1. For the dominating role of the resistance in the modern age, see for example Jean-Marie Schaeffer, *La fin de l'exception humaine* (Paris: Gallimard, 2007).
2. Steve Baker, *The Postmodern Animal* (London: Reaktion Books, 2000), 20.
3. See for example Stephen F. Eisenman, *The Cry of Nature: Art and the Making of Animal Rights* (London: Reaktion Books, 2013).
4. See Jacques Derrida, *The Animal That Therefore I Am*, trans. David Wills (New York: Fordham University Press, 2008). Derrida's term consists of the two French words denoting "animal" and "word" *(mot)*. Quite finely, the term also is able to solve the problem of reduction inherent in the process of naming all the animals with one word, since *animot* is pronounced the same way as the plural of the French word *animal (animaux)*.
5. See especially Michael Fried, *Absorption and Theatricality: Painting and Beholder in the Age of Diderot* (Chicago: Chicago University Press, 1980) and Norman Bryson, *Word and Image: French Painting of the Ancien Régime* (Cambridge: Cambridge University Press, 1981).
6. See Schaeffer, *La fin de l'exception humaine*, passim.
7. Georges Bataille, *Œuvres complètes*, tome VII (Paris: Gallimard, 1971–1988), 293.

1 The Exceptionality of the Human Spirit

> All effects of art are merely effects of nature for the person who has not attained a perception of art that is free, that is, one that is both passive and active, both swept away and reflective. Such a person behaves merely as a creature of nature and has never really experienced and appreciated art as art.
>
> —Friedrich Schelling, *Philosophy of Art* (1802–03)[1]

With regard to the concept of the animal, the modern age could be characterized as an epoch of human exception. The ideas held about humanity were, for most of preceding time, determined by the view that man's spiritual qualities are unique. One of these was man's ability to make art. Since these abilities were often contrasted to those belonging to (other) animals, the concept of animality was strangely accented. As in the quotation above, animality loomed at the margin of art theory.

In this part, I will show how this tendency to mark the territory of art in contrast to animality reappeared in different philosophical contexts. While in the latter parts of my work, I am going to present how the concept of the animal resisted this exclusion, and how animal life even could be posited as a model for the modern avant-garde, this part strives above all to point out the qualifications of this exclusion and its extent. I will first rely on Jean-Marie Schaeffer's analysis of the "thesis of human exception," which I also will try to deepen a little in connection with our topic—a thesis that Schaeffer thinks has been dominating Western views about culture ever since the time of René Descartes. Afterwards, I will clarify how the German philosophical tradition, which dominated 19th-century aesthetics and mostly presented itself as hostile towards Cartesian dualism, still suppressed the possibility of the revaluation of the concept of the animal.[2]

The Human Exception

To begin with, Schaeffer claims in *La fin de l'exception humaine* (2007) that the Occidental thesis of human exceptionality is grounded on four postulates:

Firstly (1), by positing the *ontic rupture*, following Schaeffer, the thesis of human exception always tries to maintain that the "way of being human," plain and simple, "can't be reduced to animal life *as taken as it is*."[3]

Secondly (2), the claim to *ontological dualism*—the driving force of Cartesian epistemology—asserts that there exists two ontologically separate layers. The argument does not concern itself only with the separation between animals and men, but touches also upon the integrity of the human being himself. Man becomes divided between two natures, which he himself more or less in vain tries to reconcile (animal and human, spiritual and material, etc.).[4]

Thirdly (3), the thesis of human exceptionality, in Schaeffer's terms, always postulates a *gnoseocentric viewpoint of humanity*. Most clearly, the idea can be seen appearing in such conceptions as those arguing that man takes part in the divine reason (as in Descartes), or in those presenting man's freedom as being based on his abilities to reason (most often, on his ability to be conscious of himself). Regardless of form, the gnoseocentrist thesis insists that the exclusively human characteristics can be found in the nature of man's consciousness.[5]

Finally, *the fourth postulate* (4), according to Schaeffer, makes up an *antinaturalist* stance regarding the truth value of any given argument. This postulate is the most important for Schaeffer's own work—maintaining that the thesis of human exception has led to the current segregationist attitudes that philosophers and cultural theorists hold against natural scientists and their methods—but it has also affected the field of art theory, at least from the period of Romanticism. To be sure, the postulate does not demand an invalidation of the whole construction of the given naturalist statements, whatever their character, but the statements themselves always need to be backed up by a deduction that is founded on some categorical premise leading to an "upper layer." The postulate insists that the essential characteristics of human nature, and of the exclusively human way of having knowledge, must be defined as something that surpasses the conditions of animal cognition. Consequently, the postulate does not only separate the "upper layer" from the purely cognitive means that we or other animals possess, but also proscribes their value.[6]

The foundation of the thesis of human exception, as already stated, can be found in Cartesianism. Although Descartes's thought may not carry within itself the ultimate form of the thesis, as Schaeffer tries to suggest,[7] it certainly is quite harsh with regard to the animal. In his *Discourse on the Method* (1637), Descartes notoriously based his thinking on a dualism between man and the animal, between the divine soul to which man is connected and the machine-like life of the animal, and since then his idea of the "animal-machine"[8]—so suspect already during his own time[9]—implanted itself so deeply into the Western philosophical imagination that it loomed behind our notions of animal life even into the 20th century.

Descartes's argument, about which there has been much fuss, consists of the idea that since animals do not possess immortal souls they neither exist as reasonable creatures nor have the capacity to experience consciously any sentiments (like joy or pain).[10] The view—so hard to accept—was founded on his thought that since animals do not possess incorporeal souls, they are not capable of expressing or communicating *rationally* any *active* state that would *give us any knowledge* about their subjectivity or their feelings. Thus, as animals lacked speech, there was no convincing evidence that could make Descartes believe that they ever possessed such qualities.[11] Since we have no proof of the soulful sentiments belonging to animals, according to him, we can regard them as being just like machines—"mechanical and corporeal." Descartes himself puts the thesis in the following terms, affirming the principles of ontic rupture and ontological dualism:

> [T]here are two different principles causing our movements. The first is purely mechanical and corporeal, and depends solely on the force of the spirits and the structure of our organs, and can be called the corporeal soul. The other, an incorporeal principle, is the mind or that soul which I have defined as a thinking substance. Thereupon I investigated very carefully whether the movements of animals originated from both of these principles or from one only. I soon perceived clearly that they could all originate from the corporeal and mechanical principle, and I regarded it as certain and demonstrated that we cannot at all prove the presence of a thinking soul in animals.[12]

Thus, Descartes may state that animals are like clocks that have been set mechanically to work after the intentions of their maker, God. Though humans have been produced the same way with respect to their bodies, they are different than animals, since men are created to take part in the immortal and divine soul. The only, but definitive, difference between men and God, in connection with the question of the animal, is therefore the human inability to build such complex mechanical structures as animals.[13]

It is true that Descartes's argument on human exceptionality, grounded on divine principles, may have contained nothing new to some critics of the Western attitude, since the whole Judeo-Christian tradition has taken as a given that man was made in the image of God, and highlighted that animals were innocent creatures living under the guidance of God's providence, unaffected by original sin—for the same reason excluded from the protection of the religious community, the Church.[14] The difference is in Cartesianism's strict gnoseocentric essentialism. Although the gnoseocentric viewpoint can be traced to Aristotle, who also maintained that while men had *logos*, the other animals didn't,[15] the argument that the difference is essential and not of degree becomes a cornerstone of a philosophical construction in Cartesianism. The reason is that the Cartesian belief marks a rupture between the modern philosophies and the medieval systems by imparting the idea that there is a distance between the model and the image, the medieval distance between man and God, by interiorizing the model into the image and suggesting a point of collision between the immanent and the transcendent.[16] This sort of gnoseocentric argument has been, in the 19th and in the 20th century, taken to its ultimate measures in the existentialist philosophies in which the human spirit has come to be considered free and autonomous, a subject constituted by and for itself.[17]

The gnoseocentric view of humanity is in one sense the nucleus of the whole thesis of man's exceptionality, since it makes the construction hard to attack from the inside. Traditionally, Cartesian dualism has been resisted either by more or less empirical evidence or by a subversion of its radical dualism, but neither of these attacks has had much effect on the gnoseocentrist view. "The reductionist naturalism," Schaeffer notes, "serves quite simply the reductionist strategy that Descartes had employed for the non-human world before developing it to an internal dualism living inside man," since "to reduce a pole to another there needs to be already a frame that is made out of these two poles of which the other one will be *afterwards* considered to be empty of life."[18] In truth, the "reductionist" theses could make arriving at conclusions on the differences between men and animal just as hard. Spinoza, from his position, easily concluded that "we can use animals at our pleasure," since "their affects are different in nature from human affects."[19]

The difference in the affects—or whatever the point of distinction—was almost always gnoseocentric. To take a couple of classical examples, Gottfried Wilhelm Leibniz and John Locke, both of whom considered "animal spirits"[20] to be essential for an artistic production of the feeling of "beauty and harmony"[21] nonetheless reduced these concessions to a secondary position. Leibniz, according to whom the "animal spirits" testified of a "perfection that shows itself in great freedom and power of action"[22]—and who had resisted the Cartesian thesis by admitting that animals did have souls, and that they even took part in the divine spirit[23] (or saying that they at least could appear so if only the Cartesianists could "take greater pains to teach animals"[24])—argued, in spite of this, that animals, in contrast with men, were not capable of "*apperception.*" The term signified for Leibniz a sufficient condition of rationality that included perception, sensation, and consciousness of self (animals being capable only of the first two).[25] Locke, in his turn, considered "animal spirits" in humans as being only "trains of motion"[26]—trains of "customs settling the habits of understanding and determining the will"[27]—evidently returning to the metaphor of animal mechanics of which Descartes had made extensive use.[28]

On the whole, one could almost say that Descartes's influence can most of all be proved by the resistance that it aroused. Many indeed resisted the extremity of his claims, but the counter-arguments often could not help but affirm the idea that man's life was thoroughly gnoseocentrist, a proposition leading to the fourth postulate: the true meaning of things, for man, is not in their worldly natural life, but in the way they affect man's proper consciousness.

Although the last idea is already written into Descartes's famous method, the politics of antinaturalism acquires its true meaning as a principle of the theory of the arts only under the influence of Romanticism. In its ultimate form, it also degrades the philosophical model itself, and enables one to claim the superior status of poetry and arts relative to philosophy. That is why the questions aroused by the fourth postulate play an important role in this book. An example of the definitive version of the thesis, which even tries to invalidate the philosophical language spoken by the theorist himself, can be found from the fragments of Novalis, who willingly declared the artist's intuition to be superior in understanding the truth about nature. "The poet," Novalis wrote, "understands Nature better than the man of science."[29] "Poetry is absolute truth. That is the gist of my philosophy: The more poetic, the more truthful."[30]

To continue in Schaeffer's footsteps for a moment, the combination of the four postulates has given him a reason to analyze and to name the major discourses to which the postulates have given birth, and that continue to propagate the thesis of human exception. Each of the three he identifies transcend human life beyond the sphere of the purely natural. Although the discourses are intertwined, their arguments can also oppose each other because of the different characteristics of the human exception they lean toward. Since all three will become more than familiar to us in the course of this work, I would like to introduce them:

1) The "philosophical" form of the thesis transcends—with regard to animal life—the instance that may pronounce "I," the instance of human subjectivity.[31] It is modern subjectivity that has lost the distance between the divine model and the image and has become "a place from which objects and objectivity can be established."[32] In modern art theory and aesthetics, the discourse makes itself felt most easily by equating the ability to make art with the exclusively human ability to be self-conscious.

2) The second form of the thesis "sets the place of the transcendence in the social sphere."[33] Its quest is to declare that none of the human social practices—whatever they may be—can be reduced to the natural level, since nature has already been surpassed by the means of a social construction in which the given practice can appear by and for the human mind.[34] The "biological life of a human being is," in this sense, "only a substratum."[35] In this regard, art is essentially a human practice with no connections to animal habits.

3) The third form of the thesis is evidently the most exclusive and the most narrow in its scope, but its consequences for the theory of the arts under high modernism are the most recognizable. It is based on an idea of *cultural transcendence*, not as a necessary social bond but as a particular form of symbolic activity. Although the three discourses do overlap with each other, the third form of the thesis often attempts to devalue the other two: it devalues them by showing how man can transcend the social law (and the hierarchies on which it is grounded) as well as his own identity. Thus, the discourse somehow maintains that humanity should be determined only as an abstract potentiality *to be something other than what it is assumed to be*. The surpassing of the given boundaries can be established through humanity, not only as an opposition towards nature but perhaps more importantly as an opposition against the social, understood in its stagnant, contractual state.[36] The human ability to transgress his own laws, even if the potential stays unrealized, defines in this sense man's exceptionality. At the same time, all conceptions of man's own origin will be denounced as inadequate because of the supplementary and amending nature of our culture.[37]

What is most puzzling in the last of the attitudes, often made explicit by modern artists and aestheticians, is that it is able to elevate human consciousness over every possible form of awareness, even while it defends its own antirational and antinaturalist stance. In its most absolute form, it can elevate the artist's mind even to the level of divine consciousness, for example when it declares that the artist belongs to the legacy of Prometheus.[38] Even an artist like Marcel Duchamp, who usually disconnected himself from "all investigations of this kind,"[39] and tried to keep a moderate distance on any sort of essentialism, was not immune to this form of argumentation. In a television interview given in 1956, he, with an admirably condensed way of speaking, summed up this form of the thesis:

> I like to look at the intellectual side of things, but I don't like the word "intellect." For me intellect is too dry a word, too inexpressive. I like the word "belief." In general when people say "I know," they don't know, they believe. Well, for my part, I believe that art is the only form of activity in which man, as man, shows himself to be a true individual who is capable of going beyond the animal state. Art is an outlet toward regions which are not ruled by space and time. To live is to believe, that's my belief.[40]

It is not, of course, in any way unexpected that many theorists and artists might have been ready to insist that art was an exclusively human activity. In many cases, it makes sense. What I have found surprising is the evident need to raise the question time after time in a context in which its status was ambiguous. Why in the world does one need to explicate one's ideas of the nature of the arts through some vague philosophical conceptions of the animal as the totally other? The theorists of art themselves, though they leave their remarks on the animal on the margin, seem almost obsessively to need to locate a point where animality stops and humanity begins.

Natural Aesthetics

The whole of modern tradition is marked by its equivocal liaison with the primitive and the natural. There have been times when the relation between the cultural and the natural—or between the new and the primitive—has been marked with a sort of a disequilibrium, but most often the terms needed each other to help the theorists to consolidate the best parts of each. However, in the 18th century, the conception of nature, and of the natural, had its day. From Charles Batteux's *The Fine Arts Reduced to a Single Principle* (1746), in which he openly declared that it is artist's task "to imitate nature,"[41] to Johann Joachim Winckelmann's *Reflections on the Imitation of Greek Works in Painting and Sculpture* (1755), in which he proposed a particular way of apprehending nature by assuming that if the artist lets his "hand and senses be directed by the Greek rule of beauty, the modern artist goes on the surest way to the imitation of nature,"[42] the natural was seen as essential for good art. In other words, the best minds of the century seemed to agree that good art was on the side of nature, and that art's naturality enlivened our human culture.

The underlying problem was not, however, so easily abandoned: How could one accommodate the natural and the acquired? How far could one go? Of course, "the natural" did not signify the same nature in which the animal lived, implicating on the contrary a bunch of varying human methods of achieving the impression of naturality. Would the apparently natural and the culturally acquired then equal what was truly natural? The importance of the subject was argued by many, and the answers were divided. Alexander Gottlieb Baumgarten introduced the question as the second principle of his *Aesthetica* (1750):

> § 2. The natural degree of the condition of the lower faculty of cognition, as developed only by its use, i.e. without dogmatic teaching, can be called NATURAL AESTHETICS. And, just as is the case with natural logic, one can distinguish between innate—the innate beautiful mind—and acquired aesthetics, which in turn can be divided into teaching and practicing.[43]

The real reason for bringing this topic into focus is that it plays an important role in perhaps the most influential aesthetical construction with regard to the modern theory of the arts, namely, the theory formulated by Immanuel Kant. The question is all the more relevant, since Kant's philosophy has lately experienced a revival, which has seen it heralded as a more natural view of aesthetics and art than the one proposed by high modernity. Regardless of these arguments, I would like to show the restrictions of Kant's view on the topic, at least insofar as they can be discussed within the boundaries of our subject.

First of all, Kant's revolution seems a bit twofold with regard to the concept of the animal. On the one hand, Kant is critical of Descartes and refutes the Cartesian thesis of animal-machine by claiming that "animals also act in accordance with representations."[44] Moreover, Kant even repeatedly refers to a "production of certain forms . . . which we call artistic instinct in animals."[45] On the other hand, this does not mean, for Kant, that animals would not be essentially different than human beings, and despite that he acknowledges that animals and humans can be observed in the same epistemological level "as living beings,"[46] he understands the distinction categorically by insisting on man's totally different possibilities of action. Furthermore, Kant returns to the Cartesian problem about how the instincts could be affirmed, since without a firm knowledge of the nature of the animal mind the "artistic instinct in animals" might as well as be something totally different. As a matter of fact, the Cartesian choice finally appears to be the only way to consider the question, since an animal instinct can't be in any sense, according to Kant, driven by reason:

> Thus, in comparing the artistic actions of animals with those of human beings, we conceive of the ground of the former, which we do not know, through the ground of similar effects in humans (reason), which we do know, and thus as an analogue of reason, and by that we also mean to indicate that the ground of the artistic capacity in animals, designated as instinct, is in fact specifically different from reason, but yet has a similar relation to the effect (comparing, say, construction by beavers with that by humans).—Yet from the fact that the human being uses reason in order to build, I cannot infer that the beaver must have the same sort of thing and call this an inference by means of the analogy.[47]

Despite appearances, Kant does not, in the above, return to a position from which the animal consciousness and its way of perceiving the world would be understood as a pure and impenetrable mystery. On the contrary, Kant's understanding of the distinction between men and animals can be seen as setting its boundaries more firmly than before. What is left behind in his works, in general, is the doubt that animals could have some *inner sense* that could determine their aesthetic sensibility—a condition that some theoreticians of the 18th century were willing to take into account.[48] In contrast, the Kantian refutation of the animal's inner sense, i.e., a denial of their having the intuition that drives people to create art, is centered on a categorical sorting out of the separate faculties of human consciousness—faculties of which the animal is seen as possessing only a few.

Whatever could be said of the various arguments regarding animal nature that Kant formed in the course of his studies—whether the human-animal border can be drawn based on man's capacity to represent the world, on men having an objective perceptual awareness, on men having subjective states, etc.[49]—they seem to be coherent throughout on at least two essential points. Firstly, Kant is clear on the fact (1) that animals are not able to develop *self-consciousness* (which, in Kantian terminology, is the criterion for a specific human way of having *intuition* and *inner sense*, both based on transcendence); and secondly, (2) that animals do have the same kind of faculty of imagination as human beings. While the second point, it is true, leaves a margin for the concept of the animal in the discourse on the arts—a margin we shall examine more later—the first of the propositions is now the more important to us. Before heading forward, we should nonetheless note that what makes the combination of the two propositions so startling is Kant's total exclusion of any form of consciousness from animal imagination. According to Kant, "we ascribe to these beings a faculty of sensation, reproductive imagination, etc., but all only sensible as a lower faculty, *and not connected with consciousness*;"[50] in other words, "consciousness is entirely lacking in animals, their actions happen according to laws of the power of imagination, which nature placed in them."[51] One sees clearly that animal imagination as considered here is neither active nor creative.

To go back to the first of the propositions, the argument that maintains that the ability to perceive the world self-consciously is an exclusively human one, one should note that, in Kant's works, its unique quality should be understood as determining and encompassing the whole arsenal of the human faculties of consciousness—of understanding, apperception, and of having an inner sense—as they are all grounded on the possibility of the pronouncement of the "I" (or vice versa).[52] In other words, this consequently means that apperception is ruled out for animals along with self-consciousness, and that Kant not only denies intuitive capacities to animal life, but also connects the faculty of self-consciousness to a possibility of "having an inner sense." The faculty of inner sense, by its turn, is, according to Kant, "a faculty which cannot be derived from some other faculty . . . a fundamental faculty which . . . can only belong to rational beings . . . a faculty [on which] the entire higher faculty of cognition is based."[53] This responds to his conception of the animal nature as ontologically different:

> Animals are accordingly different from human souls not in degree but rather in species; for however much animal souls increase in their sensible faculties, consciousness of their self, inner sense, still cannot be attained thereby. Even though they have better phenomena in sensibility than we do, they still lack inner sense . . . Accordingly, just as our intellectuality will increase in the other world, sensibility in animal can also increase, but they will never become equal to us. Now we can think problematically that such beings exist which have no inner sense, for it is no contradiction to suppose such. How many phenomena can be explained, without assuming an inner sense, from the faculty of outer sensibility in such beings that have no inner sense? The consciousness of one's self, the concept of the I, does not occur with such beings that have no inner sense; accordingly no non-rational animal can think: I am; from this follows the difference that beings with such a concept of the I possess *personality*.[54]

As one can see, the concept of intuition has been separated from animal spirits, which in an empiricist context equalled transcendent states of consciousness. Intuition is, in Kant's legacy, determined by the concept of self. The Kantian influence is in this sense felt through the following centuries and marks the German tradition throughout.

However, the strange alliance between intuition and a capacity to represent an "I" is mediated through a twofold concept of time in which, despite appearances, "the purely natural" becomes degraded. The historical time of the human life is connected, in Kant's understanding, to inner sense. "Time is nothing but the form of the inner sense, that is, of the intuitions of self and of our inner state,"[55] he writes. He thus emphasizes the condition of living humanely in the world: that is, having an intuition (i.e. inner sense), experiencing historical time, and being creative through the capacity to represent the human "I." This does not evidently mean that there wouldn't be in an animal's mind "nothing new under the sun," as G.W.F. Hegel later tried to suggest,[56] but it follows that the animal's relation to time is essentially different from man's. For man's apperception allows a consciousness of temporal order, i.e. an order evidenced by the self that is constructed within time, "while the broader [animal] sense corresponds merely to the capacity for temporally-ordered consciousness."[57] As is seen, Kant does not deny the animal a sort of an experience of time, nor a capacity to represent objects of its own perception, but for the reason already stated—the animal's lack of self-consciousness—the animal belongs to a pure sphere of nature, since it relies on the imagination. Or, as Colin McLear has put it, Kant thinks that for the animal its representations follow one another "according to associative regularities governing the animal's faculty of imagination."[58] It means that the animal has no experience of time in the subjective sense, since only a dialectics between historical time and "the higher order" lets the human intuition, as well as man's subjectivity, appear as that which "can be given previously to all thought."[59]

Such mental ability becomes important for our topic since it also governs, in Kant's understanding, the laws of art and the experience of beauty. It is also why Kant can explain "the art of genius" as being a consequence of the antithetical relation between imagination and self-consciousness—a dialectics purified of all animal impulses, but still a dialectics that lets oneself dive into the natural by reconciling the human "I" with nature. "Genius is," writes Kant, "the innate mental disposition through which Nature gives the rule to Art."[60] In this reconciliation, therefore, is affirmed the principle of natural aesthetics, which badly needs the thesis of man's exceptionality: "*Beautiful Art is an art, insofar as it seems like nature*,"[61] Kant concludes.

In this sense, it is also understandable why some critical commentators have thought that Kant is even more severe towards animals than Descartes. Despite Kant's allowing animals some form of subjectivity, unlike Cartesians, animal life is no longer even seen as evidence of the impressiveness of the work of the divine clockmaker, but only as an inferior form of life lacking the abilities of the human spirit. Thus, Kant's idea of the human capacity for aesthetic judgment—a capacity driven by the inner sense—has restored the locus of beauty to the human mind itself. Nowhere is this as detectable as in Kant's view of the sublime, since the human, according to Kant, differs essentially from the rest of nature because of man's ability to access a sublime experience of the natural. The fact is manifested unambiguously in his statement maintaining that experience of the sublime "reveals a capacity for judging ourselves as independent of [nature] and [gives for us] a superiority over nature on which is grounded a self-preservation of quite another kind than that which can be threatened and endangered by nature outside us."[62]

This anthropocentric sphere of the sublime is grounded in Kant's analysis, which asserts that although beauty is related to understanding, the sublime can only be experienced with the faculty of reason, i.e., with the supersensible cognitive powers belonging to humanity that are in themselves evidence of the faculty of mind by which "every standard of sense is transcended."[63] That is, everything is transcended but the essentially human—whatever that may be.

Origin and Language

A question that has haunted modern philosophy and the modern discourse on the arts from the 18th century is the one of art's own origin. Perhaps surprisingly, the question has been colored from the start by its relationship to the concept of the animal; the contemplations of the origin of language and art did not need to wait for the triumph of evolutionary theory, or the discovery of the Altamira Cave, to reach their full ambiguity regarding the human-animal boundary. Often, of course, the question was posed only to substantiate our own exceptionality.

In the latter parts of this work, we will deal with many of the theories concerning the questions raised on behalf of art's (animal) origin. For now, I would like to introduce the question in a frame of reference we have already dealt with: that in which the primacy of reason, exemplified by a human capacity to produce a certain kind of language, keeps its privileged place. Indeed, since the theoretical field insists that man is categorically different from all the other creatures as an *actively thinking being*, and since man's thinking has been considered inseparable from his ability to *produce language*, it followed that the question of language has been given an unmeasurable value with regard to the origin of art. These connections might explain the increased interest in the origin of language, art, and poetics in the wake of the Kantian-Romanticist idiom.

To be sure, the field in which the aesthetical and art historical questions on the origin of art were discussed was quite fragmented and has been ever since. In general, the discussions can be divided in accordance with their aims: they vary from (1) descriptions of the ideal moment of the natural state to (2) attempts to give an account of the historical origin of art and language over to (3) philosophical arguments in which the origin was understood as a specific human subjectivity constituting the ontological exceptionality of the species. Often, it follows that a confusion between these different objects of research takes place. Therefore, the answers also vary according to the point on which the emphasis was laid, that is, whether the theorists tried to present the origin as empirical or as a purely mental fact.[64]

The confusion between the two poles of the question has substantial consequences within a discourse on the origin of art, which has been usually organized hierarchically around the demands of the discipline of aesthetics. Following Tsion Avital's terms, this has often led to a situation in which the archeological starts to dissipate into the epistemological.[65] This is, I suggest, just what starts to happen in the historical context under discussion; oddly, the concept of the animal gets entangled in the process, and the process even affects negatively the valuation of animal art. Thus, to give an example of the possible consequences, still observable in the 20th century, the image of an animal found in an ancient artwork loses its cultural signification before the primacy of the question of the cognitive attributes that are assumed to be needed to produce a similar kind of work. Moreover, as Avital notes:

> When the cognitivist looks at the earliest painting found (assuming for the moment that there is agreement about this), the question that occupies him is not "Does

> there exist an earlier painting that has not yet been discovered?" or "From what paintings or marks did this painting evolve?" but rather "What are the cognitive attributes without which this person could not have painted this picture?" In other words, his question transfers the question of the origin from the archaeological domain to the epistemological domain, since for him the earliest possible origin of painting cannot itself be an empirical fact, but must be a cognitive one.[66]

Historically speaking, when at the end of the 18th century the question of the origin of art was raised, it almost couldn't be allowed to appear without a "cognitivist" statement made upon it. This need, I maintain, made the concept of the animal a reference point for the art theory of the era: the concept appeared as a negative term, representing the real other to whose deficiencies the possibility of creating art—a practice determined in advance as being an exclusively human one—was compared.[67]

The origin of art was now also the origin of human language: it is a common view for the theorists of the 18th and the 19th centuries that the two are of the same origin, that the first language of man was poetics. The implications of the view are manifold: it let its supporters criticize, in a manner coming down from Jean-Jacques Rousseau, the decadence of common language and its appearance in modern society, but it also served as a shield against the empiricist theories that had considered the possibility of animal speech. Evidently, the definitions of art and poetics differed from one theorist to another, but it was the animal that was finally excluded almost every time. In this sense, it becomes understandable why Johann Gottfried von Herder, to take an influential example, in his *Essay on the Origin of Language* (1772), before granting that poetry was "the natural language of all beings fashioned by reason into sounds, into images of action, passion, and living impact,"[68] had to remind his readers that "the structure and design and even the earliest cornerstone of this palace [constructed by human language] reveals humanity."[69] For Herder, man was the only one of the natural creatures capable of creating with "proportion and order."[70]

Whatever the points of reference, the differences in degree were usually seen as insufficient when the human-animal border was in question. Herder had indeed warned against the viewpoints of Rousseau (particularly those presented in the *Discourse on the Origin and Basis of Inequality Among Men*) and Étienne Bonnot de Condillac (particularly those of *Traité des animaux*), criticizing the former for making men into animals and the latter for turning animals into men.[71] For Herder, as suggested, it was the primacy of hearing and of language that enabled man to live beyond the animal state, a state in which the primacy of other senses determined the direction of life: "We creatures of hearing stand in the middle," he exclaimed, "we turn, as it were, into hearing through all our senses!"[72]

The concept of the natural becomes a problem once more, after the different mental faculties have been used to explain the origin of language. This happens, when in early Romanticism, Kant's idea of the inner sense starts to combine with a Romanticist theory of the poetics, the latter gathering all the arts underneath it: the origin of all the arts is evidently the origin of human language, the linguistic "structure and design" that comes from man's interior. While the view guarantees the primacy of poetry among the arts, it also states that the purely natural—the ground, the origin, the absolute—can only shine through in an artwork that is poetical by its nature: "There are writers who drink the absolute like water; and books in which even the dogs refer to the infinite,"[73] wrote Friedrich Schlegel. Man's artistic nature was now seen to be his true nature, and the sphere of art was claimed to be the essentially human one. As a matter of fact, this second nature was the true end of the first. "Just as a child is only a thing which wants to become a human being," Schlegel concluded, "so a poem is only a product of nature which wants to become a work of art."[74] If Herder had believed in man's unique intellectual abilities that were proven by language, and through which man could overcome the fragmentary nature of his own sensual and purely natural impressions, Friedrich Schelling thought now that the poet's privilege was to "apprehend nature spiritually."[75] As with Schlegel, Schelling's nature was of course not the nature that animals knew, but a nature that maintained a fundamental difference with regard to the "blind operations"[76] of the animals. In any case, the work of art, nature filtered through human intelligence, had become nature's highest aim.

As suggested, this was the beginning of a motion further away from the sphere of nature. One of these paths was exploited by Hegel in his works discussing the origin of art. Although Hegel's approach is altogether different than those discussed above, his understanding of art's origin follows the same path, but he goes a step further. "Man," according to Hegel's *Aesthetik*, "is separated from the animal by the infinite difference."[77] This infinity is, respectively, an essential difference between animal needs and human desire, the latter of which is structured dialectically.[78] Therefore, after Hegel, man is different than the animal "even in man's animal functions,"[79] man being, in other words, a process of the rationalization of desires leading to "self-conscious science."[80]

For this reason, Hegel is able, in his usual way, to approach the limit of the discourse by supporting the idea of the human exception and insisting that "just for the reason that [man] knows himself to be animal, man ceases to be animal, and, as *mind,* attains to self-knowledge . . . breaking the boundary of merely potential and immediate consciousness."[81] Whatever the stakes of the statement at hand, the remark on the immediateness brings us back to the problem of the origin, an origin considered as a proof of the cognitive exceptionality of man that is determined by language. According to Hegel, the man, the artist finds himself in the center of a never-ending cycle. One can only admiringly follow how Hegel explains "the ground and the necessary origin" of an artwork. An artwork, created in man's own image,[82] synthesizes the dialectics between *the inner and the outer,* the *past and the present,* and makes "knowledge" coincide with "action" in "free rationality" that is considered to be the origin of man and his arts:

> The universal need for expression in art lies, therefore, in man's rational impulse to exalt the inner and outer world into spiritual consciousness for himself, as an object in which he recognizes his own self. He satisfies the need of this spiritual freedom when he makes all that exists explicit for himself *within,* and in a corresponding way realizes this his explicit self *without,* evoking thereby, in this reduplication of himself, what is in him into vision and into knowledge for his own mind and for that of others. This is the free rationality of man, in which, as all action and knowledge, so also art has its ground and necessary origin.[83]

Hegel's historicist view that maintains that the history of art has gone through its *symbolic, classical,* and *romantic* phases includes the image of the animal in its dialectics, but only to primitivize animal imagery. He describes the progression that labels the images of animals, as representation of gods, as "abstract attributes"[84] in the symbolic phase; while in the classical phase their role has already been abandoned for the sake of a coincidence between the human and the divine that has as its output an anthropomorphic art.[85] The classical coincidence is, according to Hegel, nonetheless only "purely *immediate* and *not explicit*, and capable of adequate manifestation in an immediate and sensuous mode,"[86] but in it "the Greek god becomes the object of naïve intuition and sensuous imagination . . . therefore, the bodily shape of man."[87] This is why, for Hegel, classical art still lacks "the power of developing painting and music as absolute forms for their content,"[88] a process which will be perfected in the Romantic phase, which "takes possession of painting and music,"[89] and more importantly, understands the primacy of language and of poetry. It makes "poetry . . . conformable to all types of the beautiful, extends over them all, because the artistic imagination is its proper medium, and imagination is essential to every product that belongs to the beautiful, whatever its type may be."[90] In other words, Hegel admits that the gnoseocentric concept of art, under the overarching name of poetry, takes control of art's periodicity and makes the images of history wither into the distance, into art's end.[91]

With Hegel's theory, the "cognitivist" stance that claimed man's poetic interiority—Kant's inner sense turned into a *purely* aesthetic category—had won the battle against the "archeological" and had excluded the animal from aesthetic considerations both as subject and as object. At the same time, the image of the animal that loomed in the archeological origin had been emptied of all of its possible meanings for the sake of an art whose eternal origin was in man's self-consciousness and in language. At his most radical moments, Hegel even denied a place for the "beauty of nature" in his system: "by the . . . expression [of 'Philosophy of Fine Art'] we at once exclude the *beauty of Nature* . . . asserting that artistic beauty stands *higher* than nature."[92]

Nature Created in Man's Image

The Romantic era is nevertheless, one must remember, an era when nature was held in high esteem, and Romanticism is still appreciated for its landscape paintings. In this sense, it might come as a surprise that even the art of the landscape contributed to the degradation of the animal in German Romanticist debates on the arts.[93] Writers and artists were not, of course, altogether hostile towards animals, but the animal was not considered in any way essential to the nature that landscape paintings showed. One might wonder what kind of nature exists without animals, as was demanded from the artists? The answer, according to many Romanticists, was that such a nature is a human one by its *measures*.

The measuring of the world in human terms—and not, for instance, by divine principles—is an inheritance that Classicism passed to Romanticism: in this sense, Hegel is right. The insistence on this principle—that the measuring should be made, and art with it, within an anthropocentric frame—was an important theme in Romanticism. In human measures, art was supposed to be great, even if, as often was the case, one wanted to express an object that surpassed human limits and produced an experience of the sublime.[94] It is then no surprise that Hegel, among others, when trying to determine the "three general predicates" of an artwork in his *Aesthetik*, was so convinced of the idea that the first two of the predicates were written down in fully antinaturalist terms. According to these principles, (1) "the work of art [was] to be no natural product, but brought to pass by means of human activity," being (2) "essentially made *for* man, and, indeed, to be more or less borrowed from the sensuous and addressed to man's sense."[95] Further, Hegel continued to praise the spiritual nature of the exclusively human art, which had "been moulded in harmony with mind."[96] As a consequence, one could easily make a valuation, a hierarchy, concerning the subjects of art; as one might expect, the animal did not come out as the winner.

Of the Romantic variation of the art of landscape painting, it should be noted that it had indeed travelled far from the natural aesthetics still in bloom in Kant, who could allow an important aesthetic value for a natural landscape. What Romanticism did embrace in Kant's terms was the idea of the experience of the sublime—a "negative pleasure"[97] as Kant called it—which excluded the natural itself, despite the fact that the experience may have been catalyzed by natural imagery. Kant's idea had been that the experience of the sublime proceeded through three separate moments: in the first one, the human subject was on firm ground, in a strict relation to his object, while in the second phase his equilibrium was unbalanced.[98] Nevertheless, in the third, the experience of the sublime enforced its subjective qualities by differentiating the subject from whatever is considered natural. Following Kant, the third state of the experience of the sublime revealed a human superiority that could not be endangered.[99]

The Romanticist tradition usually emphasized this particular distance, while it saturated its own conception of the poetic with the characteristics of the sublime. It is no wonder then that Hegel, among others—in spite of his attempt to debase the Kantian sublime by marginalizing it as a subcategory of a common aesthetic experience—let his understanding of the landscape be wholly marked by it. What was perhaps new in Hegel's thought, in contrast with Kant's, was the total abandonment of the natural landscape for the sake of an art of painting that was measured on a human scale. In Hegel's own words, a landscape painting was considered a spiritual manifestation of a reason or a feeling and should respectively "*assume a higher rank than the mere natural landscape*,"[100] since the spiritual value of "human interest" could be "apprehended in the work of art, and exhibited more purely and transparently than is possible on the soil of common unartistic reality."[101]

The curious temptation to exclude the image of the animal from the picture of nature in German thinking, following the principles given above, seems to be a theme that has constantly been overlooked, although it had a huge influence on the gaze that was cast on the images of animals until the turn of the 20th century. Perhaps the fiercest guardian of the principle of art in this sense, in which even the images of animals were excluded from the sphere of true art, was Johann Wolfgang von Goethe. In his works, the ultimately narcissistic image of the artist, already hinted at above by Hegel in his analysis of the "classical phase," reaches its peak. This might be a surprise, since one has learned to consider Goethe a keen admirer of nature and an ardent investigator of her qualities. Indeed, Goethe does insist that the artist must be an enthusiastic examiner of nature; but this artist acquires praise only if he will turn his face towards nature through the principles of great art. "He to whom nature has once begun to reveal her open mystery," Goethe writes, "comes to experience an irresistible yearning for her most worthy interpreter, namely for art."[102] Thus, the target of his investigations is not the knowledge of nature, but the self-knowledge of man, since "man is the highest and the proper subject of art,"[103] and to understand him "a general knowledge of organic nature is imperative."[104]

One should note a remarkable consequence that follows from giving an intrinsic value to man's self-knowledge in artistic theory: Goethe's idea degrades all figures whose reduction to the self-knowledge of man can be made to sound unthinkable—in general, the heterogeneous other. "We," as human beings, Goethe emphasizes, "know no other world than that which stands in relation to mankind [and] we desire no other art than that which is an image of this relationship."[105] Of course, one could defend Goethe by thinking that it is the whole of nature that can be made to speak to man of himself, but Goethe is quite explicit in calling for the hierarchy of the objects of research, and, for that matter, for the subjects of art. His conclusion is much more radical than any Cartesian could make, since it leads to a conscious exclusion of all but anthropocentric ideas from the sphere of high art. "The highest aspiration of art is to present human forms in the most sensuously significant and most beautiful fashion that is possible," he writes.[106]

If the subject of the given genre nevertheless seemed to surpass the demands, as was the case with landscape painting, it should still be made to fit the measures implicated in the theory. The exclusion of the animal from the picture of nature was declared with much clarity at least by Karl Ludwig Fernow, who was hoping for artists to paint "ideal natural scenes"[107] that correspond to our inner landscapes or "representations of . . . imagination."[108] In his article "On Landscape Painting," Fernow appeared to exclude animals from landscape painting: he invoked a hierarchy that was, as a sign of the times, determined according to the quality of the intellectual ideas about humanity that could be, in one way or the other, expressed in a given artwork. "In order to determine the aesthetic status of the different arts," Fernow wrote, "we should not consider the nature of the objects they depict nor the greater or lesser difficulty which they require to master, but rather the *greater or lesser capacity for expressing aesthetic ideas* which each art possesses."[109] Such art was to combine "the sensuous and intellectual forces of the soul and of bringing them into a state of free and harmonious accord"[110]—a task for which there was not much use for the image of the animal. For the sake of clarity, Fernow's conclusion, separating the image of the animal from the noble art of landscape painting, should be quoted in its entirety:

> For this reason, it is not animal painting which should be considered next in status after history painting but landscape painting. For landscape painting is more capable of bringing the mind into an aesthetic state . . . We may concede that animal painting, as the art of depicting living creatures, requires a greater gift of imagination and more knowledge of physiognomy in order to capture the proper character of each type of animal and to articulate with truth and animation the fleeting expression of different movements and feelings. Nonetheless, we must recognize that landscape painting presupposes a higher aesthetic culture in the artist than animal painting. If the artist wishes to bring about in us a state of aesthetic harmony through his work, he must himself have felt the same concord in respect of the idea which informs it. And whoever wish to lift our souls above reality to the level of ideas must himself have cultivated his imagination in such a way that he is responsive to this domain. The strings which he seeks to sound harmoniously within us must first have sounded harmoniously within him. Such effects, which cannot be demanded from a work of animal painting, are expected from ideal landscapes.[111]

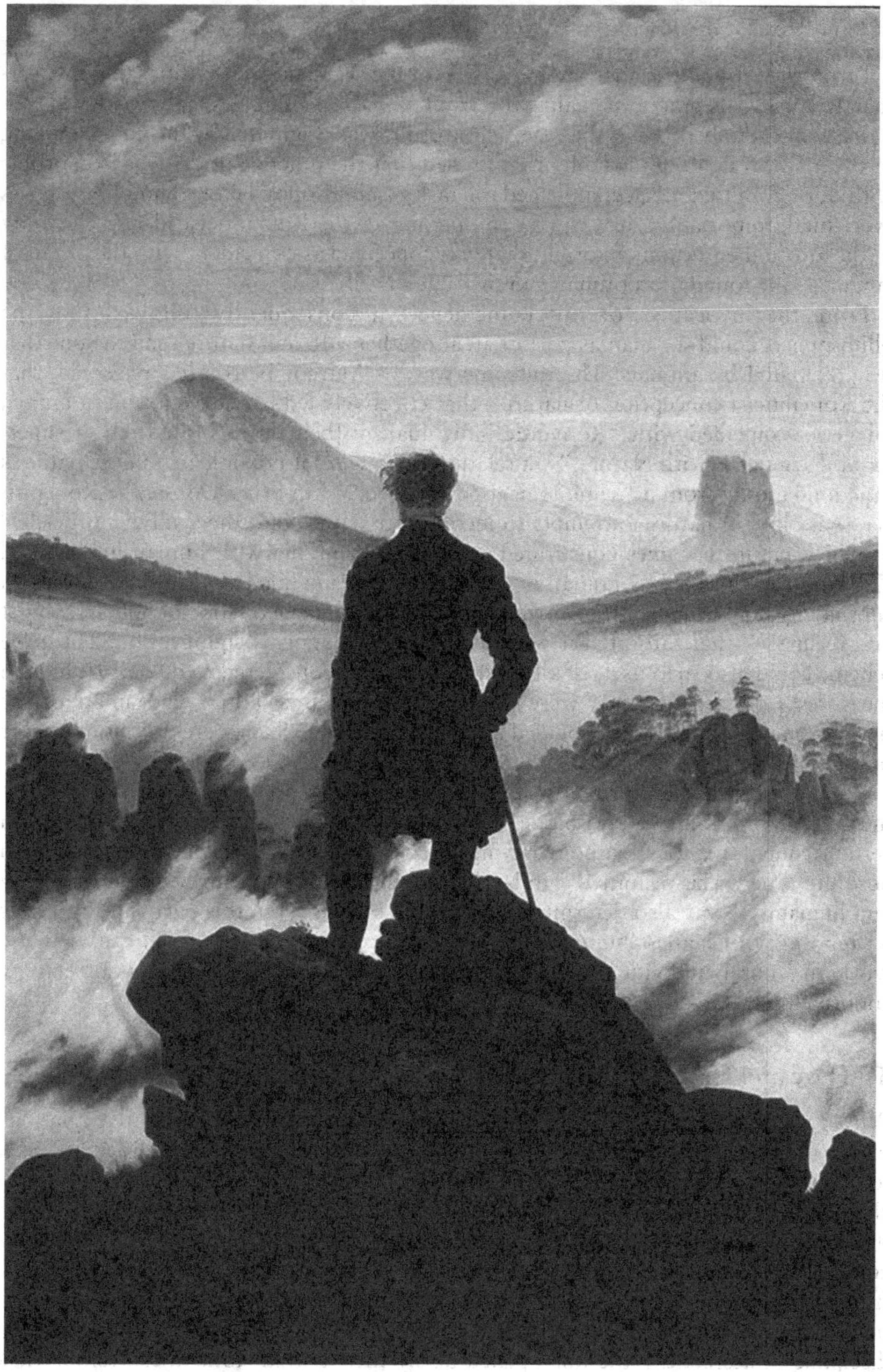

Figure 1.1 Caspar David Friedrich, *Wanderer Above the Mists*. c. 1818. Oil on canvas, 94.8 x 74.8 cm. Hamburger Kunsthalle, Hamburg. Available from: ARTstor, www.artstor.org (accessed November 7, 2016).

Fernow's testimony is an accurate version of the segregationist politics that the German Romanticists practiced against the art of *animaliers*. Of course, there were exceptions, but generally this was the case. The attitude was originated by the understanding of the arts through a philosophy that argued that the practice of art was isomorphic with it, or even that art accomplished the task of philosophy better than philosophers ever could. Importantly, it was also an outcome of the Romanticist idea of the unity of the arts in their origin: an origin that was conceived as coincident with the origin of language, the foundation of man's knowledge of self.

From the invocation of this coincidence, it consequently followed that the philosophers could see man as part of an altogether different nature than the one that was inhabited by animals. The outcome was, as Margot Norris has observed, that the Romanticist conception of nature—that effectively led to the exclusion of animal imagery—coincided with "Romantic individualism [being] elaborated into a supernatural sympathy with Nature,"—a relation that makes it possible to analyze nature's functioning only from a "semiological perspective."[112] In other words, the Romanticist sensibility of nature "attempts to incorporate nature into the symbolic order,"[113] in which nature becomes constituted as being measurable with human intelligence. The landscape marked by human measures could meanwhile be purified of all foreign subjectivity, and become "transformed only into a language or code of human conflict and feeling"[114] that saturated it with the means of linguistic signification. Thus, the scheme could also strip itself of all subjects that were considered to be outside human measures: from then on, they were empty of meaning and had no place in an artistic play depending on the scale provided by human language. Such were the antinaturalist politics of Romanticist landscape.

What could still be affirmed was the division that made possible man's nature's integration into itself the "primitive," and even the labeling of this primitiveness with certain animal qualities, without having to compromise its idea of the animal as an absolute other. The primitive mind—often used as a third state between animality and humanity—was also determined, according to Norris, in this particular frame as a "repository of highest human value"[115]: a naïf consciousness, which can view the world in a purely individual way outside of cultural constraints, but as a subjectivity confirmed by semiosis.[116]

To Have and Have Not

Why it was considered so crucial to define the practice of art by its negative relations to the animal? However complex the question may be, it sprang from a definite need to humanize artistic thinking in a particular theoretical context. The German tradition cherished the link that was supposed to unify the philosophic and the poetic, and in this kind of understanding of the arts the *poetic*—the artistic sphere most explicitly connected to language and its particular usage—determined the hierarchy of the arts. If poetics were considered an archetype of human self-consciousness, the animal was placed outside their boundaries. It is a curious fact that the major tradition in the discipline of aesthetics—the one opened up by the Kantian-Romanticist idiom—has somehow taught us to think that it is reasonable to define man's exceptionality through the spiritual qualities linked to art, or even, that art is perhaps the only guarantee of our humanity.

The idea of denying the links between art and animality remained in the forefront. In several philosophical forms to which the German tradition gave birth, it was renewed. Hegelianism was countered on its own terms in Marxism, and the dualist basis of Cartesianism was attacked in phenomenology, but all these movements usually affirmed the essential difference that marked the human-animal boundary. All these traditions constantly referred to a certain habit of art that was exclusively human, and these curiously demanded that it should also be stated as being so. Quite tellingly of the general need to discuss aesthetical questions in connection with the human-animal divide, Marx wrote in his *Economical and Philosophical Manuscripts* of 1844 that the difference between man and animal resided in the fact that "animals" were able to "produce only according to the standards and needs of the species to which they belong, while man is capable of producing according to the standards of every species and of applying to each object its inherent standard."[117] Thus, he was able to conclude that "man also produces in accordance with the laws of beauty."[118]

The situation was not different when art was defined as essentially a practice of empathy ("*Einfühlung*") or understanding. The aesthetics of *Einfühlung*, so important in the formation of the 20th-century discipline of art history—all the way from Heinrich Wölfflin to Aby Warburg, Erwin Panofsky, and Ernst Gombrich—serves as an exemplary case. The empathic relation defined art's essence, and it was understood as being the nucleus of art's practical functioning.[119] Evidently, the idea led to sophisticated theories on the dynamics involved, but it had to maintain its distance from the animal other to preserve its philosophical basis within the cultural context in which it was born. Ernst Cassirer, to give an example, demanded on these grounds that man shouldn't be considered as "*animal rationale*" but as "*animal symbolicum*."[120] The symbol and its creative function made the difference, and Cassirer was quite explicit: "Here we touch upon the crucial point in our whole problem [of defining humanity]. The difference between *propositional language* and *emotional language* is the real landmark between the human and the animal world."[121] The denial was as forceful as ever.

The most interesting developments in this sense were perhaps made under the premises of phenomenological tradition. This is a little surprising, since one could be inclined to think that while the phenomenological tradition opposes Cartesianism by denying the ontological separation between the mind and the body, it could also oppose the thesis of man's exceptionality. As a matter of fact, the divide between man and the animal became the cornerstone of phenomenological philosophy.

Edmund Husserl, to give the most obvious example, understood the world as being composed of a threefold hierarchy, according to which the lowest degree was a material thing [*Ding*] or a physical body [*Körper*]. Animals reach to the second level, since they, in addition to their bodies, also have souls [*Seele*] or psyche [*Psyche*]. What animals do not have is the community that is defined by Spirit [*Geist*], which belongs only to man. Without going into further detail, the spirituality in question is affirmed through "the Human I" [*Ich-Mensch*], which guarantees the intentional subjectivity of every human being and binds man morally to other subjectivities alike. Although the Husserlian view lets one even think anew about the conditions for communication between the different species, the foundation of all its ideas is the transcendental subjectivity that belongs to man alone. The soulful "I" that belongs to the animal is able to move the animal body (and other bodies) physically, but it can't, according to Husserl, be called "a member of a social world."[122] Pure transcendental subjectivity becomes man's characteristic feature and an absolute origin of perception by which the objects of consciousness are perceived.[123] Because of the human I, the human body is a special concern: in *The Crisis of European Sciences and Transcendental Phenomenology* Husserl separates the "physical body" from the "living body." The latter belongs to the human I, whose life is characterized as being one of "spiritual individuality," the former to the animal.[124] The gnoseocentrist worldview implied in Husserl's idea of the pure transcendental subjectivity thus affirms the animal again as the pure other, despite the knowledge acquired by the natural sciences of its time; the animal keeps its status as an other who cannot trespass into the transcendent.[125]

As Martin Heidegger is well known for his insistence that his philosophy aimed to put human life back into the context of its animality,[126] one could assume that Heidegger's theory of art would also have something to say on the relations between art and animality (or, at least, that it would not deny it). On the contrary, the human-animal border is confirmed in a usual way in his texts, perhaps at times even a little inadvertently. Thus, Heidegger maintained that animals are not able to perform any kind of human activity, like art, since human activities are always ontologically different by their origin.[127]

The Heideggerian view of the separating line between the animal and the human is famous for its idea of the difference between "having the world" and of "being poor in the world."[128] The view is, as is well-known, relative to Rainer Maria Rilke's idea of the animal gaze that looks into *the open*. "We compel even our youngest," Rilke had written in his *Duino Elegies*, "force each child to stare behind . . . not to see the openness which lies so deep within the gaze of animals."[129] In a Rilkean tone, Heidegger also declared that the animal stays "open in captivation"[130] and that its life is "that which is most difficult to think"[131]—by putting, in his turn, an emphasis on the impossibility of communication between the two ways of life, the human and the animal, despite their proximity. Moreover, the idea that the animal lives "open in captivation" puts an essential paradox in the center of the Heideggerian projection: the animal is doomed to live in "an open" in which it becomes impossible for any animal to confront its existential conditions, or to disclose them. This is how Heidegger himself conceives the difference:

> The animal possesses this being-open in its essence. Being open in captivation is the *essential possession* of the animal. On the basis of this possession it can be deprived, it can be poor, it can be determined in its being by poverty. This having is certainly *not a having of world*, but rather being held captive to the disinhibiting ring—it is *a having of that which disinhibits*. But it is because this having is a being-open for that which disinhibits, it is because the very possibility of having whatever disinhibits manifest as a being is withheld from this being-open-for, that this possession of being-open is a not-having, and indeed a not-having of world—if the manifestness of beings as such does indeed belong to the world. Accordingly we do not at all find in the animal a simultaneous having and not-having of world, but rather a *not-having of world in the having of openness for whatever disinhibits*. As a result, the not-having of world is not merely a case of having less of world in comparison with man, but rather a case of not having at all—but this now in the sense of a not-*having*, i.e., on the basis of a having.[132]

This attitude proves only to be a reformulation of the thesis in its entirety. The *ontic rupture* asserts itself with unquestionable force, while the argument leads into a vicious circle in which man's exceptionality is the only rule. As Schaeffer notes, the anti-essentialist stance with regard to our humanity affirms "that what in humanity is shown to escape animality is only revealed at the same moment when the nature of humanity is in the process of being defined."[133] In other words, man's essence is paradoxically identified as being anti-essentialism itself. The strange consequence of this sort of existentialist theory, arguing for anti-essentialism, is that their principle, which is supposed to leave doors wide open, in fact leads to stagnation of the determinations of human nature and anchors them into the animality that consequently becomes considered as a negative image of man's true essence—whether the difference is seen as residing in *freedom* and *annihilation*[134] (as in Sartre)[135] or in the *ex-sistence* belonging to *Dasein* (Heidegger).[136]

In one sense, Heidegger's idea is the culmination of the tradition that we have been following: it goes so far as to state that an inorganic piece of work, a work of art coming from the region of the spirit, can have much more life in it than an animal. This is at least the case when the topic is considered in the frame of Heidegger's ontological thinking given in *The Origin of the Work of Art* (1935–1937). In more precise terms, Heidegger states that the work of art has a semblance with man, since such work is not a "mere thing," because the human hand always has an "affinity" with an artwork.[137] A work of art, he claims, "opens up a world *and keeps it*,"[138] while animals are more like inorganic things like stones. In his own words, this is how Heidegger formulates these relations in connection with the assumedly spiritual origin of art:

> A piece of equipment, a pair of shoes for instance, when finished, is also self-contained like the mere thing, but it does not have the character of having taken shape by itself like the granite boulder. On the other hand, equipment displays an affinity with the artwork insofar as it is something produced by the human hand. However, by its self-sufficient presencing the work of art is similar rather to the mere thing which has taken shape by itself and is self-contained. Nevertheless we do not count such works among mere things.[139]

Later on:

> What does the work [of art], as work, set up? Towering up within itself, the work opens up a world and keeps it abidingly in force . . . The world worlds . . . A stone is worldless. Plant and animal likewise have no world; but they belong to the covert throng of a surrounding into which they are linked. The peasant woman, on the other hand, has a world because she dwells in the overtness of beings.[140]

Intentionally or not, Heidegger defines, as through philosophical inversion of Rilke's poetics,[141] the animal space in terms similar to those that had already colored Hegel's views of nature. Animal life is a life totally dominated by space and time and closed to itself and to others: "in Nature there happens 'nothing new under the sun,' and the multiform play of its phenomena so far induces a feeling of ennui; only in those changes which take place in the region of Spirit does anything new arise."[142] Pointedly indeed, Dominick LaCapra has argued that the Heideggerian scheme can thus present animals "as living fully within a *habitus*, functioning on the basis of an unquestioned environmental set of practices, and . . . locked within a repetition compulsion."[143] That is to say, once again, that they live in a space where there is no room for creativity, only mechanics.

*

However, that which was denied kept coming back. The thesis was often opened up to its own futilities. The surprising extent to which the definition of art was connected to the human-animal border had its own problems, not least because the theory of the avant-garde was so eager to change the nature of the arts, as we shall discuss.

What I would still like to make clear is that the critics of the modern attitude towards animals—even those who limited themselves to the discussions on the arts—are totally right in one sense: almost always, when man's exceptionality was contested in a modernist frame, it was done through the terms to which the thesis of man's exceptionality itself was grounded. This also had its peculiarities. While the references to the animal world had become so ordinary, the animal also took on an ambiguous status, understanding of which was seen as essential in knowing what art was. Not only, then, did the animal stand out as the privileged other contrasted with the human habits, but the theorists also quickly noticed how ambiguous its value was. It even could mean that, from time to time, the animal could be used to mark a space of apparent resistance. "A fragment, like a miniature work of art," Friedrich Schlegel observed, "has to be entirely isolated from the surrounding world and be complete in itself like a porcupine."[144]

Notes

1. Schelling cited in Robert M. Pippin, *After the Beautiful: Hegel and the Philosophy of Pictorial Modernism* (Chicago: University of Chicago Press: 2014), 1.
2. This introduction does not mean to say that there haven't been animal lovers and theoreticians in the modern and early modern ages who have struggled against the restrictions laid down by the thesis, but the theoretical tradition, as I hope to show, has been dominated by the disposition that proposes that man has exceptional spiritual qualities in contrast with animality, and that the same qualities make up a necessary condition for making art.
3. Jean-Marie Schaeffer, *La fin de l'exception humaine* (Paris: Gallimard, 2007), 27.
4. Ibid., 27–28.
5. Ibid., 28.
6. Ibid., 28–29. One could be critical towards Schaeffer's claim that the thesis implies an anti-naturalistic standpoint, at least when concerned—in the context of this research—with some of the theories of art. One should notice that the character of "the natural," which has often been given a significant value on its own, is almost always affirmed only if it is discernable that "the naturality" is mediated through human consciousness. We shall be discussing the topic later on at length.
7. Ibid., 48–137.
8. See René Descartes, *Œuvres de Descartes*, tome VI (Paris: J. Vrin 1996), 55–58.
9. The best known of the critical commentaries given by Descartes's contemporaries was written by Pierre Gassendi. On the polemics, see René Descartes, *Oeuvres de Descartes,* tome VII (Paris: J. Vrin, 1996), 262–272.
10. See Descartes, *Oeuvres de Descartes,* tome VI, 55–58.
11. Ibid, 55–58.
12. René Descartes, *Philosophical Works of Descartes,* vol. III, ed. John Cottingham, Robert Stoothoff, Dugald Murdoch & Anthony Kenny (Cambridge: Cambridge University Press, 1991), 365.

13. Descartes, *Oeuvres de Descartes,* tome VI, 55–56; Descartes, *Oeuvres de Descartes,* tome V, 277. As puzzling as Descartes's statement is, there are at least three traditions in interpreting it:

    1) The statement can be taken in all its obscenity: animals do not possess immortal souls, reason, or any kinds of sentiments.
    2) The statement can be seen as valid *only* on the epistemological level, implying that since we have no sustainable proof—expressed by means of human rationality, i.e. in any human language—of the nature of the sentiments of animals, we are not able to make any statements in their defense. Thus, if there is animal subjectivity, it becomes unknowable (i.e. non-existent) because of the Cartesian imperatives of doubt. In this sense, it is maintained that Descartes does not posit his statements within phenomenality but within epistemology. See e.g. Katherine Morris, "Bêtes-machines," in *Descartes' Natural Philosophy,* ed. Stephen Gaukroger, John A. Schuster & John Sutton (London: Routledge, 2000), 409.
    3) The idea of the animal-machine should be seen only as a symptom of Cartesian dualism; it is there only to support the hierarchy descending from God through humans down to the beastly creatures. The idea has been defended, for example, by pointing out that Descartes might have meant that "animals do have sensations but they are not conscious of it" (Markus Wild, *Die anthropologische Differenz. Der Geist der Tiere in der frühen Neuzeit bei Montaigne, Descartes und Hume* [Berlin: Walter de Gruyter, 2006], 173). Of course, one could ask Wild, as Kreeta Ranki suggests, whether unconscious sensations are sensations at all. (Kreeta Ranki *Eläinten hyvinvointi uuden ajan alun eläinkäsitysten valossa* [The Welfare of Animals in the Light of the Ideas about the Animals in the Beginning of the Modern Era] [Turku: Turun yliopisto, 2010], 37).

14. See, for example, Schaeffer, *La fin de l'exception humaine*, 44–45.
15. Aristotle thought that animals did not possess "speech" (as *logos*), though they had "voices," which were but an indication of pleasure and pain. See Gary Steiner, *Anthropocentrism and Its Discontents: The Moral Status of Animals in the History of Western Philosophy* (Pittsburgh, PA: University of Pittsburgh Press, 2005), 61; originally, in Aristotle *The Complete Works of Aristotle*, vol. 2, ed. Jonathan Barnes (Princeton: Princeton University Press, 1995), §1253a. For a summary of Aristotle's ambivalent attitude towards the topic, see, for example, Steiner, *Anthropocentrism and Its Discontents*, 53–76.
16. Schaeffer, *La fin de l'exception humaine*, 44.
17. Ibid., 49.
18. Ibid., 208.
19. Despite the fact that animals, according to Spinoza's reasoning, do have sentiments, he says that when it comes to animals we may "consider them at our own advantage, use them at our pleasure, and treat them as is most convenient for us. For they do not agree in nature with us, and their affects are different in nature from human affects" (Benedict de Spinoza, *Complete Works*, vol. 4, trans. Samuel Shirley [Cambridge: Hackett, 2002, §37]).
20. "Animal spirits," a term often found in the vocabulary of early modern philosophy, mostly appears in contexts in which it is supposed to connect us with the rest of the animal kingdom through some activity, i.e., in this case, the production of artistic harmony. Descartes himself made use of the term by defining it, in his *Treatise of Man* (1648), as being "a certain quite subtle wind or, rather, a very lively and very pure flame, to which the name *animal spirits* have been given" (Descartes, *Oeuvres de Descartes,* tome XI, 120).
21. On Leibniz's view, see Gottfried Wilhelm Leibniz, *Philosophical Papers and Letters,* ed. Leroy E. Loemker (Dordrecht: Reidel, 1969), 697–699 and Gottfried Wilhelm Leibniz, *Philosophical Writings*, ed. G. H. R. Parkinson (London: Dent, 1973), 196; on Locke's, see John Locke, *An Essay Concerning Human Understanding,* vol. I, accessed May 15, 2014, www.gutenberg.org/cache/epub/10615/pg10615-images.html, XXXIII §6.
22. Leibniz, *Philosophical Papers and Letters*, 698.
23. Gottfried Wilhelm Leibniz, *Sämtliche Schriften und Briefe*, vol. 6 (Berlin: Akademie Verlag, 1923), 1678. "Angels are beings composed of spirit and body, as are men, beasts and plants."
24. Ibid., 1474.

25. On the topic at large, see, for example Murray Miles, "Leibniz on Apperception and Animal Souls," in *Dialogue*, vol. 33 (Fall 1994): 701–724. Interestingly indeed, Leibniz thinks it important to show the principle of "divine wisdom" in contrast with Spinoza-like reductionism, while still resisting the Cartesian temptation in Cartesianism's own terms: "I do not say that the corporeal world is a machine or a watch which goes on without God's *interposition*, and I am insistent enough that created things stand in need of His continual influence. But I do maintain that it is a watch which goes without needing His *correction*: otherwise we should have to admit that God keeps improving upon His work. God has foreseen everything, He has provided a remedy for everything in advance. There is in his works an already pre-established harmony and beauty" (Leibniz, *Philosophical Writings*, 196).
26. Locke, *An Essay Concerning Human Understanding*, vol. I, XXXIII §6.
27. Ibid.
28. In spite of the fact that Locke has many reservations about Descartes's thoughts on animals, the following paragraph should leave no doubt about the argument made in my text: "The case is not so much different in brutes but that any one may hence see what makes an animal and continues it the same. Something we have like this in machines, and may serve to illustrate it. For example, what is a watch? It is plain it is nothing but a fit organization or construction of parts to a certain end, which, when a sufficient force is added to it, it is capable to attain. If we would suppose this machine one continued body, all whose organized parts were repaired, increased, or diminished by a constant addition or separation of insensible parts, with one common life, we should have something very much like the body of an animal; with this difference, That, in an animal the fitness of the organization, and the motion wherein life consists, begin together, the motion coming from within; but in machines the force coming sensibly from without, is often away when the organ is in order, and well fitted to receive it" (Ibid., XXVII §6).
29. Novalis "Fugitive Thoughts," in *Art in Theory, 1648–1815: An Anthology of Changing Ideas*, ed. Charles Harrison, Paul Wood & Jason Gaiger (Oxford: Blackwell Publishing, 2000), 910.
30. Ibid., 911.
31. Schaeffer, *La fin de l'exception humaine*, 14. The form of the thesis is dominated by the Kantian, the phenomenological, and the hermeneutical traditions—all of which Schaeffer criticizes at length in his book.
32. Ibid., 24.
33. Ibid., 14.
34. Ibid.
35. Ibid.
36. See ibid., 15. In spite of his critique of the strictly segregationist attitude of the discourse in question, Schaeffer defends its arguments by noting that the oppositional disposition between nature and culture can't be unraveled without a separation of the cultural from the social (ibid., 221). As Peter Bürger has noted, Kant's idea of the autonomy of an aesthetic judgment was used in Romanticism to define the whole sphere of the arts. This, according to Bürger, made possible the idea that art—all art—has been allowed to represent the revolutionary side of humanity through the idea that the human subject is able to overcome the boundaries of the natural and the communal. Understandably, Bürger criticizes this view for forgetting its own historicity. (See Bürger, Peter, *Theory of the Avant-Garde*, trans. Michael Shaw [Minneapolis: University of Minnesota Press, 1989], 46.)
37. With the preceding analysis, Schaeffer has mainly tried to point out that an essentializing attitude towards the human-animal divide has contributed to the segregationist process in the scientific field. In this process, the humanistic disciplines have separated themselves from the new results acquired by the natural sciences. After the rise of Romanticism, according to Schaeffer, this development seems to have been reinforced, and the postmodernist turn, which at least in some cases seems to have been able to resist the thesis of human exception, has not been able to lessen the segregationist process. Schaeffer continues by noting that in spite of the support for the doctrine of interdisciplinarity, the current theoretical field has only confirmed the assumedly higher status of philosophical theory with regard to other discourses, with philosophy often supporting itself with claims that it is driven by some form of metascientific approach. (Schaeffer, *La fin de l'exception humaine*, 14–24 & passim.)

38. On such ideas, see for example Georges Bataille, *Œuvres complètes*, tome 1 (Paris: Gallimard, 1971–1988), 497–500.
39. Duchamp wrote in a letter to André Breton—on Michel Carrouges's analysis of the atheist character of his work *La mariée mis à nu par ses célibataires, même*—commenting on his attitudes towards questions of belief and on the existence of God: "For me," Duchamp confessed, "there is something other than *yes*, *no*, and *indifferent*—it is for example the *absence of investigations of this kind*." (Marcel Duchamp, *Affectt Marcel*: *The Selected Correspondence of Marcel Duchamp*, ed. Francis M. Naumann & Hector Obalk [London: Thames and Hudson, 2000], 343.) The statement is often understood as condemning the whole field of metaphysics, and is interpreted as a refusal to take any sort of essentialist standpoints. (For more details, see for example Calvin Tomkins, *Duchamp: A Biography* [New York: The Museum of Modern Art, 2014], 387.) The same conclusion can be made also in the light of another famous citation from Duchamp: "All this twaddle," Duchamp writes, "the existence of God, atheism, determinism, free will, societies, death, etc., are the pieces of a chess game called language, and they are only amusing if one does not preoccupy oneself with winning or losing this game of chess" (Ibid., 390). After Baudrillard's era, one is of course familiar with the conclusions that can be made from Duchamp's agnostic position: when all knowing is reduced to believing, or to a game, one might as well start to believe in fiction.
40. Ibid., 389–390.
41. Charles Batteux, "The Fine Arts Reduced to a Single Principle," in *The Bloomsbury Anthology of Aesthetics*, ed. Joseph Tanke & Colin McQuillan (New York: Bloomsbury, 2012), 140–158; see, for example 151.
42. Johann Joachim Winckelmann, *Reflections on the Imitation of Greek Works in Painting and Sculpture* (London: A. Millan, 1765), 19, accessed May 15, 2014, https://archive.org/details/reflectionsonpai00winc.
43. Alexander Gottlieb Baumgarten, "Aesthetica," in *The Bloomsbury Anthology of Aesthetics*, ed. Joseph Tanke & Colin McQuillan (New York: Bloomsbury, 2012), 158.
44. Immanuel Kant, *Critique of the Power of Judgment*, ed. Paul Guyer (Cambridge: Cambridge University Press, 2000), p. 328n §90.
45. Ibid., p. 308 §85.
46. Ibid., p. 328n §90. Note that this argument is above all a confirmation of the ontological dualism that conditions the thesis of man's exceptionality.
47. Ibid.
48. One interesting example comes from Francis Hutcheson's classic text *An Inquiry into the Original of Our Ideas of Beauty and Virtue*. Hutcheson writes, "We seem scarce capable of answering such Question [concerning God] any way; nor need we enquire, whether other Animals may not discern Uniformity and Regularity in Objects which escape our Observation, and may not perhaps have their Senses constituted so as to perceive Beauty, from the same Foundation which we do, in Objects which our Senses are not fitted to examine or compare." Francis Hutcheson, "An Inquiry into the Original of Our Ideas of Beauty and Virtue," in *The Bloomsbury Anthology of Aesthetics*, ed. Joseph Tanke & Colin McQuillan (New York: Bloomsbury, 2012), 130.
49. On these determinations, for example Colin McLear, "Kant on Animal Consciousness," in *Philosopher's Imprint*, vol. 11 no. 15 (November 2011).
50. Immanuel Kant, *Lectures on Metaphysics*, ed. Karl Ameriks & Steve Naragon (Cambridge: Cambridge University Press, 1997), 88. My emphasis.
51. Kant, *Lectures on Metaphysics*, 390.
52. See e.g. McLear "Kant on Animal Consciousness," 8.
53. Ibid., 10.
54. Kant, *Lectures on Metaphysics*, 87.
55. Immanuel Kant, *The Critique of Pure Reason*, trans. J. M. D. Meiklejohn (Salt Lake City, UT: Project Gutenberg, 2003), Intr. §7, accessed May 15, 2014, www.gutenberg.org/files/4280/4280-h/4280-h.htm. Translation modified.
56. Georg Wilhelm Friedrich Hegel, *Philosophy of History*, trans. J. Sibree (Marxists Internet Archive), §60, accessed May 15, 2014, www.marxists.org/reference/archive/hegel/works/hi/index.htm.

57. McLear "Kant on Animal Consciousness," 11.
58. Ibid., 14.
59. Immanuel Kant, *The Critique of Pure Reason*, II. §12.
60. Immanuel Kant, *Critique of Judgment*, trans. James Creed Meredith (Mineola, NY: Dover, 2012), p. 112 §46. The statement is given in a paragraph called "Beautiful art is the art of genius" in Kant's 3rd critique.
61. Ibid., p. 111 §45. My emphasis.
62. Kant, *Critique of the Power of Judgment*, p. 145 §28.
63. Kant, *Critique of Judgment*, p. 66 §25. Translation modified.
64. On the subject, see, for example Tsion Avital, "The Origins of Art: An Archaeological or a Philosophical Problem?" in *South African Journal of Art History*, vol. 16 (2001), 4.
65. Ibid., 6.
66. Ibid.
67. See more on this topic in Part 4.
68. Johann Gottfried von Herder, "Essay on the Origin of Language," in *On the Origin of Language*, trans. John H. Moran & Alexander Gode (Chicago: University of Chicago Press, 1986), 135–136.
69. Ibid., 131.
70. Ibid., 147.
71. Ibid., 103.
72. Ibid., 143. Only a few pages earlier, Herder comments on sheep's bleating as a form of language. In any case, this has been done only to highlight that human language is the most effective way of surpassing the dominion of the other senses: "There is for instance the sheep. As an image it looms before the eye with all things and images and colors on a great canvas of nature. How much is there and how difficult to distinguish! All the characteristic marks are finely interwoven, placed together, and all still ineffable! Who can speak shapes? Who can sound colors? Let him take the sheep under his probing hand. This sensation is more secure and fuller, but it is so full and so obscure, with one thing within the other. Who can say what he is thus feeling? But listen! The sheep bleats! Now one distinguishing mark separates by itself from the canvas of the colors wherein so little was to be distinguished. One distinguishing mark has penetrated deeply and clearly into the soul. 'Oh,' says the learning beginner, like Cheselden's blind man when given the power of sight, 'now I shall know you again—you bleat!'" (Ibid., 129.)
73. Friedrich Schlegel, *Friedrich Schlegel's Lucinde and the Fragments*, trans. Peter Wirchow (Minneapolis: University of Minnesota Press, 1971), p. 149 §54.
74. Ibid., p. 145 §21.
75. Friedrich Schelling, "Concerning the Relation of the Plastic Arts to Nature," in *Art in Theory, 1648–1815: An Anthology of Changing Ideas*, ed. Charles Harrison, Paul Wood, & Jason Gaiger, trans. Michael Bullock (Oxford: Blackwell Publishing, 2000), 938. "In all natural beings the living idea is manifested in blind operation only: if it were the same in the artist he would differ in no way from nature. If, however, he were consciously to subordinate himself entirely to nature and reproduce the existent with servile fidelity he would produce masks, but no works of art. Thus, he must withdraw from the product or creature, but only in order to raise himself to the level of creative energy and apprehend it spiritually."
76. Ibid.
77. Georg Wilhelm Friedrich Hegel, *Introductory Lectures on Aesthetics*, trans. Bernard Bosanquet (London: Penguin Books, 2004), 86. Translation modified.
78. The formulation refers to Alexandre Kojève's interpretation of Hegel in which he states that the Hegelian dialectic is "not a *method* of research, but the adequate description of the *structure* of Being, and of the realization and appearance of Being as well" (Alexandre Kojève, *Introduction to the Reading of Hegel*, trans. James H. Nichols Jr. [Ithaca, NY: Cornell University Press, 1980], 259).
79. Hegel, *Introductory Lectures on Aesthetics*, 86. Hegel adds: "[—] as, for instance, [in] the process of digestion."
80. Ibid.
81. Ibid. Translation modified.

82. See ibid., 84–85.
83. Ibid., 36.
84. Ibid., 83.
85. Ibid., 82–88.
86. Ibid., 86.
87. Ibid. Translation modified.
88. Ibid., 97.
89. Ibid.
90. Ibid. As we have already seen what kind of ideas Hegel (as well as Kant) had on the differences between the time experienced by the animal and the historical time that belonged to man, it is interesting to note that the discourse in which the different arts were evaluated in their various relations to time and space (a central concern in German art theory ever after Lessing) was also affirmed by anti-Hegelian thinker Søren Kierkegaard. Kierkegaard had written on the topic that "If one traces dialectically and just as much historically the development of the aesthetically beautiful, one will find that the direction of this movement is from spatial categories to temporal categories, and that the perfecting of art is contingent upon the possibility of gradually detaching itself more and more from space and aiming toward time. This constitutes the transition and the significance of the transition from sculpture to painting, as Schelling early pointed out. Music has time as its element but has no continuance in time; its significance is the continual vanishing in time; it sounds in time, but it also fades and has no continuance. Ultimately poetry is highest of all the arts and therefore also the art that best knows how to affirm the meaning of time. It does not need to limit itself to the moment in the sense that painting does; neither does it disappear without a trace in the sense that music does. But despite all this, it, too, is compelled, as we have seen, to concentrate in the moment. It has, therefore, its limitation and cannot, as shown above, portray that of which the truth is precisely the temporal sequence. And yet this, that time is affirmed, is not a disparagement of the aesthetic; on the contrary, the more this occurs, the richer and fuller the aesthetic ideal becomes" (Søren Kierkegaard, *Kierkegaard's Writings IV, Part II: Either/Or*, trans. Howard W. Hong & Edna H. Hong [Princeton, NJ: Princeton University Press, 2013], 136–137). The primacy of the poetic arts, or the reduction of every art to its "poetics," is also a crucial term when one is concerned about the status of the animal other in the midst of the discourse on the arts, since it confirms that what was needed from the arts was a relation to the world that could be seen as situated as far as possible from the animal sphere—Kierkegaard continues his analysis with an overall criticism of the aesthetic sphere by demanding an aesthetics to "be lived" instead of an "aesthetical portrayal in poetic reproduction" (Ibid., 137). As we shall later see, the pictorial arts were considered by some critics as offering an experience dangerously close to experiencing the world through the eyes of an animal.
91. See e.g. Hegel, *Introductory Lectures on Aesthetics*, 46–61.
92. Ibid., 3–4.
93. As we shall see later, the French variation of the Romanticist discourse, influential among the visual arts, could more easily make statements favorable to the concept of the animal.
94. The idea that the sublime is *something that exceeds human measures or man's capacities of cognition* had already been famously asserted by Edmund Burke in the middle of the 18th century. (See Edmund Burke, *A Philosophical Enquiry into the Origin of our Ideas of the Sublime and Beautiful* [Oxford: Oxford University Press, 1990], passim.) But as I shall later argue (in Part 3), Burke's views on the sublime were posited in a way that could lead to a totally different form of aesthetics. His views, more remarkably, allowed an important place for the animal with relation to the experience of the sublime.

    Moreover, the 18th-century English tradition was interested in the sublime in a somewhat different way than the Kantian-Romanticist tradition. Jonathan Richardson had already understood the sublime in its relations with an artwork in a sense that could be described as being wholly modern in 1725. Richardson had written in his *Essay on the Theory of Painting*: "When we propose only an exact Imitation of Nature we shall certainly fall short of it; So when we aim no higher than what we find in any One, or more Masters, we shall never reach their Excellence: He that would rise to the Sublime must form an Idea of something beyond all we have yet seen; or which Art, or Nature has yet produc'd . . . when all the Excellencies of the several Masters are United, and their several Defects avoided." (Jonathan Richardson, *An Essay on the Theory of Painting* [London: A.C., 1725], 259–260, accessed May 15, 2014, https://ia802304.us.archive.org/35/items/essayontheoryofp00rich/essayontheoryofp00rich.pdf.) If the ending of the passage cited from Richardson still did not allow for the sentiment of modern art, the view was later corrected and reflected by Burke, who maintained in his treatise on the sublime that "art can never give the rules that make an art" (Burke, *A Philosophical Enquiry into the Origin of our Ideas of the Sublime and Beautiful*, 49).

95. Hegel, *Introductory Lectures on Aesthetics*, 30.
96. Ibid., 33.
97. Kant, *Critique of the Power of Judgment*, p. 129 §23. It is quite illuminating to see Kant's expression compared to Burke's idea of the sublime as a "delightful horror" (Edmund Burke, *A Philosophical Enquiry into the Origin of our Ideas of the Sublime and Beautiful*, 67; see also for the comparison, Peter Heymans, "An Ecocritical Approach to the Discourse of the Sublime in 'The Ancient Mariner'," in *The Coleridge Bulletin: The Journal of the Friends of Coleridge New Series* 30 (Winter 2007): 18).
98. See, for example Heymans "An Ecocritical Approach to the Discourse of the Sublime in 'The Ancient Mariner'," 18.
99. Kant, *Critique of the Power of Judgment*, p. 145 §28.
100. Hegel, *Introductory Lectures on Aesthetics*, 34.
101. Ibid., 34.
102. Johann Wolfgang von Goethe, "Maxims and Reflections," in *Art in Theory, 1815–1900: An Anthology of Changing Ideas*, ed. Charles Harrison, Paul Wood & Jason Gaiger, trans. Nicholas Walker (Oxford: Blackwell Publishing, 1998), 75.
103. Johann Wolfgang von Goethe, *Goethe on Art*, ed. John Gage (Berkeley: University of California Press, 1980), 6.
104. Ibid.
105. Goethe, "Maxims and Reflections," 75.
106. Ibid.
107. Karl Ludwig Fernow, "On Landscape Painting," in *Art in Theory, 1648–1815: An Anthology of Changing Ideas*, ed. Charles Harrison, Paul Wood & Jason Gaiger, trans. Jason Gaiger (Oxford: Blackwell Publishing, 2000), 1067.
108. Ibid., 1067–1068.
109. Ibid., 1067.
110. Ibid., 1068.
111. Ibid.
112. Margot Norris, *Beasts of the Modern Imagination* (Baltimore, MD: Johns Hopkins University Press, 1985), 235.
113. Ibid.
114. Ibid.
115. Ibid.
116. Ibid. Norris gives as examples of the sort of primitive Shelley's prototype (and female anti-type), Keats's poet, and Rousseau's *naif*, but I have no reason to doubt that the statements could be used in connection with the context in question.
117. Karl Marx, *Economic & Philosophic Manuscripts of 1844*, trans. Martin Milligan, revised by Dirk J. Struik (Marxists Internet Archive), 32, accessed May 15, 2014, www.marxists.org/archive/marx/works/download/pdf/Economic-Philosophic-Manuscripts-1844.pdf.
118. Ibid. Although a fraction of the neo-Marxist tradition has tried to maintain that Marx's materialism allows humans to share a common nature with animals, and to use Marxist dialectics for pro-animal purposes, it must be noted that Marx was fully supporting the rhetorical figure inherited from Romanticism and Hegelianism with regard to our question (see, for example Katherine Perlo, "Marxism and the Underdog," in *Society & Animals* 10:3 [2002]: 303–318). A larger section of the text, from which the above citation has been extracted, makes Marx's position quite clear: "Man is a species-being, not

only because in practice and in theory he adopts the species (his own as well as those of other things) as his object, but—and this is only another way of expressing it—also because he treats himself as the actual, living species; because he treats himself as a *universal* and therefore a free being . . . The animal is immediately one with its life activity. It does not distinguish itself from it. It is its *life activity*. Man makes his life activity itself the object of his will and of his consciousness. He has conscious life activity. It is not a determination with which he directly merges. Conscious life activity distinguishes man immediately from animal life activity. It is just because of this that he is a species-being. Or it is only because he is a species-being that he is a conscious being, i.e., that his own life is an object for him. Only because of that is his activity free activity. Estranged labor reverses the relationship, so that it is just because man is a conscious being that he makes his life activity, his essential being, a mere means to his *existence* . . . In creating a *world of objects* by his personal activity, in his work upon inorganic nature, man proves himself a conscious species-being, i.e., as a being that treats the species as his own essential being, or that treats itself as a species-being. Admittedly animals also produce. They build themselves nests, dwellings, like the bees, beavers, ants, etc. But an animal only produces what it immediately needs for itself or its young. It produces one-sidedly, whilst man produces universally. It produces only under the dominion of immediate physical need, whilst man produces even when he is free from physical need and only truly produces in freedom therefrom. An animal produces only itself, whilst man reproduces the whole of nature. An animal's product belongs immediately to its physical body, whilst man freely confronts his product. An animal forms only in accordance with the standard and the need of the species to which it belongs, whilst man knows how to produce in accordance with the standard of every species, and knows how to apply everywhere the inherent standard to the object. Man therefore also forms objects in accordance with the laws of beauty . . . It is just in his work upon the objective world, therefore, that man really proves himself to be a *species-being*. This production is his active species-life. Through this production, nature appears as *his* work and his reality. The object of labor is, therefore, the *objectification of man's species-life*: for he duplicates himself not only, as in consciousness, intellectually, but also actively, in reality, and therefore he sees himself in a world that he has created. In tearing away from man the object of his production, therefore, estranged labor tears from him his *species-life*, his real objectivity as a member of the species and transforms his advantage over animals into the disadvantage that his inorganic body, nature, is taken from him." (Marx, *Economic & Philosophic Manuscripts of 1844*, 31–32.)

119. The idea is perhaps written down in its most paradigmatic form in Leo Tolstoy's *What is Art?* (1897). Tolstoy writes that "The activity of art is based on the fact that a man receiving through his sense of hearing or sight another man's expression of feeling, is capable of experiencing the emotion which moved the man who expressed it." (Leo Tolstoy, *What is Art? What is Religion?* trans. Aylmer Maude, V. Tchertkoff & A. C. Fifield [Milford, PA: Wildside Press, 2008], 41.)
120. See e.g. Ernst Cassirer, *An Essay on Man: an Introduction to the Philosophy of Human Culture* (Garden City, NY: Yale University Press, 1944), 44–62.
121. Ibid., 48.
122. See Barry Smith & David Woodruff Smith, ed. *The Cambridge Companion to Husserl* (Cambridge: Cambridge University Press, 1995), 351–353. Citation is taken from Edmund Husserl, *Collected Works, Vol. Three: Ideas Pertaining to a Pure Phenomenology and to a Phenomenological Philosophy*, trans. Richard Rojcewicz and André Schuwer (London: Kluwer Academic Publishers, 2009), p. 184 §49, accessed May 15, 2014, https://ia801402.us.archive.org/20/items/IdeasPartIi/Husserl-IdeasIi.pdf.
123. At least after 1907, after Husserl's so-called "transcendental turn." In relation to the topic at hand, see Schaeffer, *La fin de l'exception humaine*, 23–25 & 411n1.
124. Husserl cited in Barry Smith, *The Cambridge Companion to Husserl*, 352. Husserl's formulation of man's twofold relation to the body is just an example of the endless diversions and dualities that the phenomenological tradition inscribes into its theories on the body. One could only think of Maurice Merleau-Ponty, describing how the experience of the "flesh" is

irreducible to the idea of the "body." (See for example Maurice Merleau-Ponty, *La Nature: Notes du Collège de France* [Paris: Éditions du Seuil, 1994]; on the topic in connection with the thesis of human exception, see Schaeffer, *La fin de l'exception humaine*, 128.)

125. What is striking in Husserl's account of the human-animal divide is the criticism that he is able to generate through its principles. As human exceptionality is assured from the moment when the hypothesis of the auto-constitutive nature of human consciousness becomes approved, it means that Husserl may connect it with his, in Schaeffer's terms, "antinaturalist politics." The following citation should leave no doubts of the nature of his attempt: "From a moral-practical standpoint, I am treating a human being as a mere thing if I do not take him as a person related to the moral, as a member of a moral association of persons in which the world of morals is constituted . . . Again, analogously, I treat a human theoretically as a thing if I do not insert him in the association of persons with reference to which we are subjects of a common surrounding world . . . He who sees everywhere only nature, nature in the sense of, and, as it were, through the eyes of, natural science, is precisely blind to the spiritual sphere, the special domain of the human sciences [*Geisteswissenschaften*]. Such a one does not see persons and does not see the Objects which depend for their sense on personal accomplishments, i.e., Objects of 'culture'." (Husserl cited in Smith, *The Cambridge Companion to Husserl*, 354; Husserl, *Collected Works*, vol. 3, p. 200–201 §51.)
126. Heidegger has been considered by some thinkers to be a predecessor of posthumanism, since he, in his "Letter on Humanism," criticizes philosophical humanism for elevating the human over the animal. See Martin Heidegger, *The Basic Writings*, ed. David Farrell Krell (New York: HarperCollins, 2008), 213–266.
127. As we may see, the fact is flagrantly manifested in the Heideggerian concept of human "ex-sistence" determining the *Dasein*.
128. See Martin Heidegger, *The Fundamental Concepts of Metaphysics: World, Finitude, Solitude*, trans. William McNeill & Nicholas Walker (Indianapolis: Indiana University Press 1995), 186–274 & passim.
129. Rainer Maria Rilke, *Duino Elegies*, trans. Stephen Cohn (Manchester: Carcanet, 1989), 65. On Rilke and the animal, see also Part 4.
130. Heidegger, *The Fundamental Concepts of Metaphysics: World, Finitude, Solitude*, 253.
131. Martin Heidegger, *The Basic Writings*, 220.
132. Heidegger, *The Fundamental Concepts of Metaphysics: World, Finitude, Solitude*, 269–270.
133. Schaeffer, *La fin de l'exception humaine*, 198.
134. *Néantisation.*
135. See e.g. Jean-Paul Sartre, *Being and Nothingness: The Principal Text of Modern Existentialism*, trans. Hazel E. Barnes (New York: Washington Square Press), 1992, passim.
136. See e.g. Martin Heidegger, *Being and Time*, trans. Joan Stambaugh (New York: State University of New York Press, 1996), passim.
137. Heidegger, *The Basic Writings*, 154.
138. Ibid., 169. My emphasis.
139. Ibid., 154.
140. Ibid., 169–170.
141. According to Giorgio Agamben's humorous formulation, it follows for Heidegger that "not only does *animalitas* become utterly unfamiliar and appear as 'that which is most difficult to think', but *humanitas* also appears as something ungraspable and absent, suspended as it is between a 'not-being-able-to-remain' and a 'not-being-able-to-leave-its-place'." See Giorgio Agamben, *The Open: Man and Animal*, trans. Kevin Attell (Stanford, CA: Stanford University Press, 2004), 50–51.
142. Hegel, *Philosophy of History*, §60.
143. Dominick LaCapra, *History and Its Limits: Human, Animal, Violence* (London: Cornell University Press, 2009), 168n27.
144. Schlegel, *Friedrich Schlegel's* Lucinde *and* the Fragments, p. 189 §206.

2 The Animal and the Image

> Our gardens are adorned with statues and our galleries with pictures. What would you imagine these masterpieces of art, thus exhibited to public admiration, represent? The great men, who have defended their country, or the still greater men who have enriched it by their virtues? Far from it. They are the images of every perversion of heart and mind, carefully selected from ancient mythology, and presented to the early curiosity of our children, doubtless that they may have before their eyes the representations of vicious actions, even before they are able to read.[1] . . . It is thus that the dissolution of morals, the necessary consequence of luxury, brings with it in its turn the corruption of taste. Further, if by chance there be found among men of average ability, an individual with enough strength of mind to refuse to comply with the spirit of the age, and to debase himself by puerile productions, his lot will be hard. He will die in indigence and oblivion. This is not so much a prediction, as a fact already confirmed by experience! Yes, Carle and Pierre Vanloo, the time is already come when your pencils, destined to increase the majesty of our temples by sublime and holy images, must fall from your hands, or else be prostituted to adorn the panels of a coach with lascivious paintings. And you, inimitable Pigal, rival of Phidias and Praxiteles, whose chisel the ancients would have employed to carve them gods, whose images almost excuse their idolatry in our eyes; even your hand must condescend to fashion the belly of an ape, or else remain idle.[2]
>
> —Jean-Jacques Rousseau, *A Discourse on the Moral Effects of the Arts and Sciences* (1750)

In this part we will deal with the developments that took place in mid-18th century France. My argument is that the French way of understanding the status of the animal, when considering the pictorial arts, took a different theoretical turn than the one allowed by the German developments. Evidently, there were connections between the French and the German thinkers, but in spite of the continual dialogue between the French essayist tradition and the most prominent German thinkers of the era, the French tradition was capable of maintaining some distance from the German.[3] One of the characteristics of the tradition, as I shall argue, is the peculiar connection of the animality and the image.

Introduction: Discourse and Imagicity

While trying to make sense of the developments in 18th-century France, we shall start with the role of visuality. In developing the theme, my aim is first of all to point out the unique way in which the qualities of *discursivity* and *imagicity* were politicized and how opposition to old political currents—an opposition linked to visuality—gained an important role within the French theories of art of the era. In this short introduction (and in the concluding note that follows), my intention is not to say that the whole field of French 18th-century art was characterized by this internal resistance based on the opposition between discursivity and imagicity, but to open, quite coarsely, the field in which the concept of the animal was implemented into the new discourse of the art that started to prevail. The opposition is prominently a French phenomenon, as it was characterized by the power of the *Académie*, whose politics were represented by the hegemony of the allegorical image.

The court of the "Sun King" has become well-known for its scopic nature; the power structure that was affirmed in the society of the spectacle linked to the cult of Louis XIV was grounded on visual exposure, a cult of outward visual appearance to which individuals submitted themselves.[4] At the same time, visuality served the structure, a structure that was mainly indexical in nature. A hierarchy of the pictorial genres, defined forcefully by the *Académie*, ratified the official nature of this arrangement: the status enjoyed by history painting and allegory was, in one way or the other, backed by an idea about a separation between the inside and the outside of the signifying process. In other words, the status of an image was dependent on its literarity, which wasn't relevant to all possible interpretations of the work, however literal, but tied to the formal characteristics of the presentation, the discursive frame in which these interpretations could become encrypted.[5] Thus, the allegorical picture was related to the power of the monarch and of the *Académie*; the idea of the allegory, relative to which the power of the sovereign was the ultimate signified, was to give the illusion, or at least to pursue the illusion, that the picture plane was ruled by the process of signification and that it could be interpreted in its entirety through its literary or indexical elements.[6] The most famous example of an artist and a theoretician of art working in this frame was Charles LeBrun, who, in his studies, tried famously to produce a vocabulary of images and gestures.

There also loomed a problem that was written inside the paradigm. As Norman Bryson has noted, the disposition,[7] which pitted the visual field, in its ambiguity, against literary elements had been implanted into early modern art ever since the introduction of the Albertian perspective.[8] The idea of the perspectival picture plane as a window to an outer world had already opened the way for a description of nature that could be seen as supplementary in relation to the principal figures of the given image, figures to which the medieval images had still been almost completely confined.[9] Thus, the task which the perspective was supposed to execute planted an ambiguity and a certain form of resistance in the visual regime, or at least, in the way the regime of the image, the ambiguity of visual perception itself, was considered.

The resistance that took place in 18th-century France launched a process in which, in Bryson's terms, "figurality" became liberated "from the controlling grasp of the signified,"[10] the "lyrical prose" made way for the "poetry" of the picture plane.[11] Bryson sees that the innovation made by Antoine Watteau played a central role in this development. While with LeBrun, the work was determined by the effort to make the signifier correspond with the signified, with Watteau, following Bryson, the signified was no longer "anchored in the precise visual signifier";[12] the works were presented more as depictions of moods than as allegories.[13] It is under this sort of formal frame that

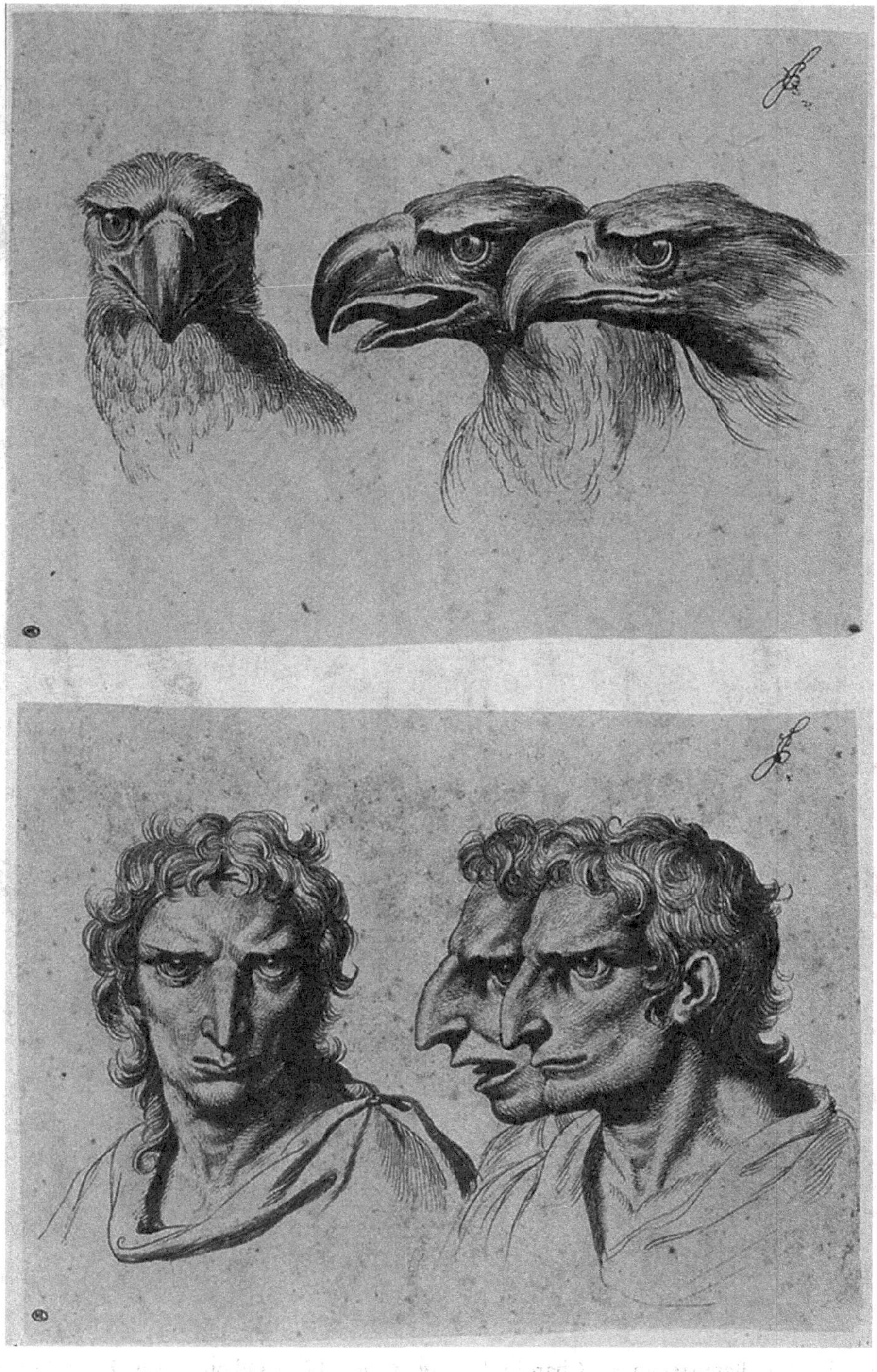

Figure 2.1 Charles LeBrun, *Three Eagle's Heads / Three Physiognomic Heads Based on the Eagle*. 17th century. Pen and ink. Musée du Louvre, Paris. Available from: ARTstor, www.artstor.org (accessed November 7, 2016).

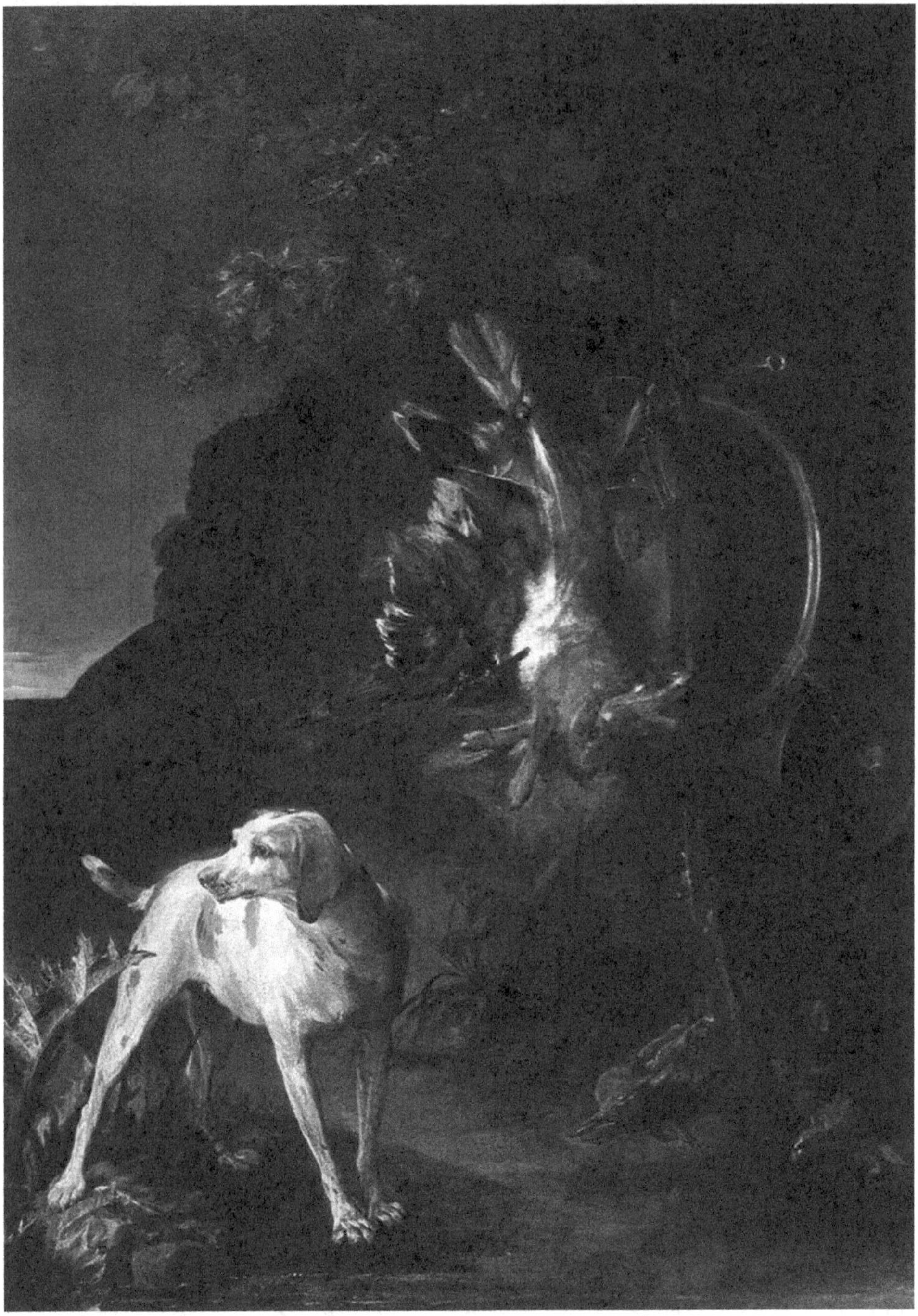

Figure 2.2 Jean-Baptiste-Siméon Chardin, *Dog and Game*. 1730. Oil on canvas, 172 x 139 cm. Norton Simon Museum, Pasadena, CA. Available from: https://commons.wikimedia.org/wiki/File:Jean_Siméon_Chardin_-_Game_Still-Life_with_Hunting_Dog_-_WGA04743.jpg (accessed November 7, 2016).

the "labour of interpretation that *knew* it was interpretation"[14] little by little lost its power, and to this development was added, in the first half of the 18th century, the birth of public art criticism, which moved the centering of the question towards value judgments and empiricist ideas. However, this is the field from which "naïve, child-like"[15] (Denis Diderot's term) art, constituting itself as public practice and conscious of itself as specific medium, sprang forth. The most telling example of this sort of resistance may perhaps be the enthusiasm that Denis Diderot felt for Jean-Siméon Chardin's works, an enthusiasm to which we shall return. "There is something which should not, with Chardin, be left unsaid," Bryson writes, "French painting will never again be so conscious of its life as *paint* and at the same time so untroubled by the consequences of that independent life."[16]

This "naïve seeing" that Diderot defended of course did not equal pure seeing in the same sense in which the late 19th-century criticism could think of it, and was not in any way purified of its linguistic nature. Instead, it constituted imagicity as only a different field of semiosis—as something that resisted transparency. However, while still not speaking the language of abstraction, it used ideological implications and erected an oppositional structure in the middle of the artistic process: then, it was possible to note, when necessary, that at least there was an area, a visual field, that could be understood as something not (yet) touched by the hierarchical structure of human speech—an ambiguous origin, as some observed. It is within this disposition that animality also started to acquire its peculiar value.

Condillac and Animal Imagination

This said, it is time to go back to Herder's complaint about Rousseau and Condillac being guilty of diminishing the watershed between man and animal. Rousseau, Herder lamented, had turned men into animals, after Condillac had already turned animals into men.[17] Although Herder overstated the situation, the remark tells us that something had happened with regard to the theorizations of the human-animal border in France. Rousseau and Condillac, to be sure, had not denied the human-animal divide altogether, but they had both considered anew the differences between human and animal ways of perceiving the world.

It really seems that Condillac's initiative was somehow essential, or at least that his distinctive empiricism had some resonance on the field. Although there were more radical thinkers in mid-18th century France, who tried to re-evaluate the human-animal interface as well as the old ideas about man's divine nature—Voltaire and Julien Offray de la Mettrie should be named—it was Condillac who appeared to be the most influential. Rousseau and Diderot were both affected by Condillac's views on the animal, and they both carried the implications of Condillac's theory to their own ideas concerning the human culture at large.[18]

Of what does Condillac's radicality consist? Ultimately the value of his arguments seems to reside in his ability to examine the problem of the human-animal boundary without disqualifying the boundary itself. In other words, he still affirms the uniqueness of human culture but denies the "infinite distance"[19] between man and animal. While his arguments involve the usual questions about the differing intellectual skills of animals and men, his texts have an exceptional emphasis on the functions of memory, language, and imagination in animal life.

This does not mean to say that Condillac would not consider that there is a hierarchy involved.[20] Although he opposes the Cartesianist claim that animals do not have souls, he states in his *Essay on the Origin of Human Knowledge* that their "souls are inferior to ours."[21] As Condillac puts it, while "the operations of the animal soul are limited to perception, consciousness, attention, reminiscence, and to an imagination that is not at its command, our soul possesses additional operations."[22] As expected, these additional operations include man's capacity to develop a much larger theoretical understanding than any animal ever can, the ability to make free decisions, to recognize the existence of God, to know the truth, to create arts and sciences, etc.[23] These differences are, he argues, mostly a consequence of the animal condition of "having only an imagination that its soul cannot direct."[24] "They," Condillac maintains, "represent things absent to themselves only to the extent that their mental image of it is closely connected with a present object."[25]

So far, nothing new, or so it seems at first glance. The situation becomes more complicated when it is understood that Condillac tries to maintain that none of the qualities mentioned above mark an essential difference between men and animals: both are governed by the same natural laws. These laws, furthermore, are based on sensations that determine a dialectic between needs and habits: "Regardless of the system of the different faculties . . . animal, in any case, is part of the general system that envelopes all the living creatures: the faculties are born of the same origin, sensation; and they are engendered by the same principle, the need."[26] What is essential for Condillac's vision, is the mutual basis on which the mental operations are founded. There are differences, and animals don't make art,[27] but they have linguistic abilities; they make use of a "language of action."[28] More importantly, the mental faculties of men and animals operate by the same means: by connecting ideas to one another.[29]

These statements leave the question of the inferiority of the animal soul open. Also, they posit a question on the intersections between the different species. What became more compelling for Condillac's followers, I argue for the time being, was the centering of the question around the concept of imagination: for Condillac, language can be either a "language of action" (of which the animals are capable), or an "articulated language" of human speech (which mediates needs and ideas to another subject), but both are always based on the faculty of imagination, which governs the mental operations involved. This is the great divide between Condillac's followers and the idiom put forward by Kant. Moreover, it is only because Condillac grants to animals these imaginative faculties that he may argue that animals "compare, judge, have ideas and a memory."[30] He states that it wouldn't make sense to deny that animals have a capacity to judge, since by sensing things they without doubt are able to compare a memory image and their present perception.[31] However, according to Condillac's understanding, the animals, not being free, are wholly dominated by the images. This condition of being dominated soon helped to relate the state of animality to the experience of the visual arts. An important development in this sense was opened up in Rousseau's texts in the 1750s and 1760s.

Rousseau and the Noble Visual

Jean-Jacques Rousseau's valuations of the animal are from the start equivocal. While Rousseau's primitivism praises qualities that refer to animality, and to a life lived in harmony with nature, he appreciates them only insofar as they belong to an exclusively human sphere of sociability.[32]

In his *"Second Discourse"* of 1754, known as the *Discourse on the Origin and Basis of Inequality Among Men*, Rousseau's views on animality are clearly already influenced by Condillac.[33] Although Rousseau makes clear that he thinks that there exists only a "difference in degree"[34] between man and animal, he centers the question around the exclusively human capacity to act as "a free agent." The paragraph in which Rousseau make this clear implies a current disturbance shaking up 18th-century French discourse on the animal. Rousseau connects himself with the fashionable problematics of the animal and the machine by putting himself between Descartes and de la Mettrie:

> I see nothing in any animal but an ingenious machine, to which nature hath given senses to wind itself up, and to guard itself, to a certain degree, against anything that might tend to disorder or destroy it. I perceive exactly the same things in the human machine, with this difference, that in the operations of the brute, nature is the sole agent, whereas man has some share in his own operations, in his character as a free agent[35] . . . It is particularly in his consciousness of this liberty that the spirituality of his soul is displayed. For physics may explain, in some measure, the mechanism of the senses and the formation of ideas; but in the power of willing or rather of choosing, and in the feeling of this power, nothing is to be found but acts which are purely spiritual and wholly inexplicable by the laws of mechanism.[36]

Thus, the animal soul is considered inferior, since it is not able to choose freely. The condition of freedom, further on, equals an ability to form "general ideas." "The only goods he [the animal] recognizes in the universe are food, a female, and sleep: the only evils he fears are pain and hunger . . . for no animal can know what it is to die; the knowledge of death and its terrors being one of the first acquisitions made by man in departing from an animal state."[37] As one already suspects, in Rousseau's mind this all gets much more complicated. The animal soul, according to him, is limited to mere sensations and memory images[38] and is not open to self-improvement[39]—this much he had already learned from Condillac[40]—but from this argument Rousseau immediately molds an opposition between the ability to form ideas through words, an ability separating animals and humans, and the condition of seeing the world as an image:

> Add to this, that general ideas cannot be introduced into the mind without the assistance of *words* . . . This is one of the reasons why animals cannot form such ideas, or ever acquire that capacity for self-improvement which depends on them. When a monkey goes from one nut to another, are we to conceive that he entertains any general idea of that kind of fruit, and compares its archetype with the two individual nuts? Assuredly he does not; but the sight of one of these nuts *recalls to his memory the sensations which he received from the other*, and his *eyes*, being modified after a certain manner, give information to the palate of the modification it is about to receive.[41]

These statements become ambiguous only when posited in the context of Rousseau's primitivism, which attacks the decadence brought by human language. Rousseau idealizes the primitive human society established at the dawn of languages, which he thinks lived according to natural laws, while in written language, on the contrary, there seems to be something that wrenches men out of nature's bosom.[42] "You will be surprised to find," Rousseau writes in *Émile*, "that I reckon the study of languages among the useless lumber of education."[43] This attitude, grounded on the somewhat futile opposition between nature and language, becomes all the more complex in the light of Rousseau's remarks in *Essay on the Origin of Languages*,[44] a book that made him go back to language's origins in his quest for the natural state. Speech, indeed, might be natural, in its ideal state—"speech, being the first social institution, owes its form to natural causes alone"[45]—but for Rousseau, the argument always returns to the system of criticism against the deficiencies of the written word. That criticism is consistent with Rousseau's twofold critique of man's ability to improve himself by the mediation of his consciousness, an ability that also predisposes man and his culture to decadence.

"The first language," man's primitive form of communication, is thus what Rousseau appreciates. This is not, it must be repeated, a purely natural condition. Without a society and without a need to feel empathy for another, he emphasizes, man "is bound to be a ferocious animal . . . having no rule but that of force," believing men to be "each other's enemies."[46] A Hobbesian war, perhaps? All this changes when common language is invented, since, according to Rousseau, it places man in a structure through which one's own passions, one's own inner language, is supplanted with another.

The original language, the inner language before the supplementation, might not be language in its proper sense, but it is a language nonetheless. Rousseau remarks in *Émile* that since the child "has only one language" it appears that it has no language at all,[47] and the process in which the child starts to understand the world through language proper is a moment when "one language supplants another."[48] Thus, "when once Émile has said, 'It hurts me,' it will take a very sharp pain to make him cry."[49] This loss is a work of empathy put forward through imaginary identification. The language "that supplants another" is the language of the other: "Although pity is native to human heart," Rousseau states, "it would remain eternally quiescent unless it were activated by imagination . . . then getting us outside ourselves and identifying us with a being who suffers."[50] The idealized language, the first social bond, is, then, for the lack of a language proper and its hierarchies, a language that speaks of another but that is bound to stay gestural and visual: "The first language had to be figurative,"[51] Rousseau concludes. It is the birth of language in imagination—figurative before it has a proper meaning[52]—that he is after.

The figurative language—the first of the figurative languages, though without appearing to be language at all—is then understood as being the language of passions. With a remarkable ease, Rousseau formulates an idea of a language that can exist without proper meanings, but that can be, within his idea, seen to be in continuation with the meanings still to be born.[53] According to him, figuration is something that may transfer meanings, since it can "transpose" the passions that it represents.[54] Nevertheless, transposition is still passionate, not rationalized. To speak of the passionate language, "as man's first motives for speaking were of the passions," is thus, imagistic as it is, to "express the passions with tropes."[55] These tropes, figurative as they were, lead to a language that in Rousseau's view first tried to "give every single word the sense of a whole proposition."[56] Figurative language, then, was a language that was by its nature poetical.

After Rousseau, the poetics appearing with this sort of passionate language are not comparable to the classical model of poetry based on metaphorical language. This is an important point, since it puts Rousseau in the middle of the cultural politics of his time, and he is not shy about the fact. One starts speaking metaphorically, he writes, only after one has lost his natural passions and started speaking according to the laws of a dead language and of common truths:

> Upon meeting others, a savage man will initially be frightened. Because of his fear he sees the others as bigger and stronger than himself. He calls them *giants*. After many experiences, he recognizes that these so-called giants are neither bigger nor stronger than he. Their stature does not approach the idea he had initially attached to the word giant. So he invents another name common to them and to him, such as the name *man*, for example, and leaves *giant* to the false object that had impressed him during his illusion. That is how the figurative word is born before the literal word, when our gaze is held in passionate fascination; and how it is that the first idea it conveys to us is not that of the truth . . . The illusory image presented by passion is the first to appear, and the language that corresponded to it was also the first invented. It subsequently became metaphorical when the enlightened spirit, recognizing its first error, used expressions only with those passions that had produced them.[57]

From here on, Rousseau astonishingly invents a common ground for the visual arts and the animal. That is to say, for the animal and the visual language; of the first language of man, that still connected man to the mythical world, and that spoke of man's inner natural passions, of his "loves,"[58] and made man to "invent drawing."[59] These connections are made to resonate with Rousseau's own critique of modern civilization. As we have seen, the critique opposes the conventional hierarchical language of man with the first figurative language that speaks of the passions. What does not go unnoticed in the *Essay on the Origin of Languages* is that his point leaves a margin for the concept of the animal. While writing, in a tone that indicates his premises, that "conventional language is characteristic of man alone, and that is why man makes progress, whether for good or ill, and animals do not,"[60] he can't miss the idea that animal life at the same time has become a model of the "passionate fascination" on which the figurative language is based. Indeed, he remarks, that if the animals do not use language proper, it might only be due to the fact that they have decided altogether not to use their ability to speak![61] However, in Rousseau's scheme, their means of communication can only be figural:

> Animals have a more than adequate structure for such communication, but none of them has ever made use of it. This seems to me a quite characteristic difference. That those animals which live and work in common, such as beavers, ants, bees, have some natural language for communicating among themselves, I would not question. Still, the speech of beavers and ants is apparently by gesture; i.e., it is only visual.[62]

An important aspect of Rousseau's views should now be taken into account: the essence of this visuality, leading up to the art of drawing, is *mimesis*. Only imitation raises the arts over the utilitarian sphere of rational language. This means that the rationalized conception of art, supporting the simplest form of the thesis of human exception, becomes untenable. These qualifications are not restricted to the pictorial arts.[63] The principle of *mimesis* is thought to connect the different arts and to let them be seen as oppositional practices against modern science: "Music is no more the art of combining sounds to please the ear," Rousseau notes, "than painting is the art of combining colors to please the eye. If there were no more to it than that, they would both be natural sciences rather than fine arts."[64] In mimetic activity, there is always a poetic supplement at stake, a promise of a true imitation and an opening towards the other that makes the activity itself ethically valuable, while natural sciences may speak without a need for empathy.

From this arises a problem we have already encountered above: the principles of empathy and *mimesis*, by their turn, introduce a hierarchy. The more related the given arts are to the principle of time, the more they give reason to reflect that the imitative signal is coming from a subject like the one receiving the signal, and the more they are understood to be "human arts."[65] This is why poetics and music are appreciated: "the one of the arts is more interesting than the other precisely because it does more to relate man to man and always gives us some idea of our kind."[66] On the same basis, Rousseau may open this discourse towards the animalized sphere of the visual, which he understands as residing in the origin of poetic language. "It is evident," he writes, "*that painting is closer to nature and that music is more dependent on human art.*"[67] These remarks should always be read along with the double structure to which they belong: they do not only affirm that the visual arts are animalized in contrast with poetics and music, but perhaps more importantly, they stress that, regardless of where the interest of the beholder lies, speech can be considered natural only from the point of its emergence, that is to say, from its figural origin. Rousseau is quite clear on this: "Man imitates, as do the beasts. The love of imitating comes from well-regulated nature; in society it becomes a vice."[68]

What Rousseau had done was to contest the boundaries of human language by representing them in the context of language's assumedly natural state. From this contestation emerged two essential oppositions that have been proved to be so important to the French theory of art ever since: the one between (hierarchic) language and (natural) speech, and the other between linguistic representations and visual gestures. These oppositions were nonetheless contrived by Rousseau from the idea that if the decadent social realities emerging from modern societies were the consequence of hierarchically structured language, this state of affairs could be made to encounter its origin in natural visual gestures. Thus, all language was determined by this internal dualism: human language contained a resistance towards language's objectifying hierarchical power, a revolutionary potential that could be linked conceptually with Rousseau's ideas of the primitive. While Rousseau obviously still maintained that there were many differences between the production of animal and human communication, he showed how the concept of the animal could be connected with the sort of "primitive" visual expression—a connection much in use with the 19th- and 20th-century avant-garde developments with which we will deal later.[69]

From Rousseau's stance, the faculty of *mimesis*, now also openly animalized, came momentarily to be seen as the original form of natural self-expression and empathy, an expression only later mutilated by a representational hierarchy in human history: "in society it becomes a vice."

Diderot's Suspicion

Rousseau did not think that animals made art, nor did he propose an animalistic conception of the human imagination. What he did open up was a conceptual connection between the concept of the animal and a communication that was thought to be predominantly visual. At the least, he implied a possibility that the visual arts could be in some way animalized.

This might seem to be a trivial move; but it had its consequences concerning the theory of the visual arts. Rousseau may not have brought the animal to the center of the discourse on the arts, but he nonetheless showed the way to do so. The particular visual sphere that Rousseau managed more or less to animalize was brought to the fore in another way in Denis Diderot's critiques.

Through and through, Diderot's politics of art are colored by his suspicion towards language. The suspicion might be relative, but it follows Rousseau's lead. "I fall into error at every moment," Diderot had written, "because language does not furnish me with the means of expressing the truth. I have one thing in the depth of my heart, I say something else."[70] While human language is "limited," says Diderot, the natural sphere of expressions is "infinite" but "mutually understood."[71]

Diderot's suspicion, which to a large extent prescribes his critical preferences also in relation to art, is anchored in his views of the false and superficial nature of human self-consciousness. His contestation of the human-animal divide, with his monist and extreme materialist proposals, is best known from his ideas put forward in texts like *Rêve d'Alembert* (1769) and *Conversation with d'Alembert* (1769). In them, Diderot often approached views that may retrospectively seem consistent with 19th-century evolutionary theories. "Then there's only one substance in the universe, in man and in animals. The bird-organ is made of wood, and man is made of flesh. The canary is made of flesh, and the musician is made of flesh organized differently but the two of them have the same origin, the same formation, the same functions, and the same end," Diderot wrote as part of his dialogues.[72] In such a world, there was no room for essential differences.[73]

Anyhow, the human condition, which seems largely determined by human language and which had often been understood through an ability to be self-aware, kept Diderot busy. In his texts, he attacked the idea of man's exceptional self-conscious nature strategically from two directions. First, by suggesting that animals might be able to be self-conscious, Diderot maintained that each "sentient being" can have retained "consciousness of having been himself from the first moment of reflection until the present moment" and that this consciousness is founded on the "memory of their actions."[74] Secondly, he went on to deny that the human mind itself could be characterized by the consciousness that it keeps of itself. This is done, in *Rêve d'Alembert*, with a simple but sophisticated gesture:

> The fibre[75] is a simple animal; the human being is a complex animal. But let's keep that issue for another occasion. To remove from a human being his consciousness

> of himself doesn't require sudden decrepitude—a considerably smaller event can do it. A man on the point of death receives the sacraments with a profound piety. He confesses his sins, asks his wife's forgiveness, kisses his children, summons his friends, speaks to his doctor, gives instructions to his servants, dictates his last wishes, puts his affairs in order—and all that with the soundest judgment, with his intelligence fully engaged. Then, once cured, he convalesces and hasn't the least idea of what he said or did during his illness. This period of time, sometimes very long, has disappeared from his life. There are even examples of people who have resumed the conversation or the action which the sudden attack of illness interrupted.[76]

The idea of self-consciousness, then, posits a problem for Diderot that holds within itself a question of the truthfulness of any given presentation. The main problem is that Diderot had lost his faith in the 17th-century dream of a language that can exteriorize its contents and remain transparent; in other words, of a language of whose contents the subject could be wholly and self-consciously aware.[77] The problem is, as d'Alembert puts it in Diderot's *Conversation with d'Alembert*, that "one can think only about one thing at a time,"[78] which further led Diderot to suggest an asymmetrical relation between intellectuality and sensitivity. In this sense, he kept alive a dream of an immanent language, an idea that also influenced his views on art. Diderot demanded that the ambiguities of the process of signification—"the symbolic fullness of the signifier," as Kenneth Berri puts it—must be brought forward against the lazy readings that determine our relations to the given symbol.[79]

That is why the question of false self-consciousness became so important. For its sake, Diderot seemed to have created his theory of the poetic "hieroglyph"—a poetic sign that does not lose its visuality, or that has a visuality at its basis. The theory is introduced already in *Lettre sur les sourds et muets* of 1751, in which the hieroglyph is considered incompatible with any vulgar notion of temporality and instead shows itself as an image, as being painted, whatever the medium through which the hieroglyph is conceived. Diderot explained the idea by saying that the hieroglyph represents the "state of the soul in one instant."[80] In other words, he referred to a poetic image that may as well as be painted with paint or molded with words.[81] The hieroglyphic moment Diderot seeks is that which exists, according to Bryson, "when the sign ceases to seem to be a sign and instead presents itself as a place of relay of unmediated presence and representation."[82] Thus, the poetic medium loses its evident value, while the notion shifts over to an area in which the qualities inherent in the process of perceiving visually are emphasized in contrast to false human self-consciousness, based on the hierarchical premises of written language.

Evidently, the idea was soon transposed to Diderot's views on pictorial arts. The means to achieve the effect can perhaps best be demonstrated through Diderot's vision of Chardin's art. The virtue of Chardin's works, for Diderot, seemed at first hand to reside in the fact that they at the same time were "natural," while they presented themselves as mediated:[83] "*Chardin est entre la nature et l'art,*"[84] Diderot wrote. He favored the naïve side of Chardin's art, of the interplay between "nature and art:" the process in which the appearance of the plane, through which the image presented itself as an illusion, did not hinder the person standing in front of the picture from becoming lost in it.

Diderot's preferences towards absorptive effects have been famously analyzed by Michael Fried, who has suggested that Diderot's aim was to establish a politics of "absorption," a condition that was considered opposite to the "theatrical." With "theatrical art" Diderot seems to have meant the pompous forms of academic art that relied on the idea of the transparency of the symbol, an art that contained an illusion that the image was made *for* the self-conscious subject totally cut off from the world of the picture. Diderot indeed concluded that if the actions and expressions of the painted figures did show any kind of attention to the beholder's presence, it interfered with the task he himself considered worthy of the art of painting. Thus, Diderot, in his turn, found himself paradoxically attached to Chardin's paintings, in which the human subjects were depicted as absorbed in their own work; the absorption of the beholder was achieved by the work's showing no signs of the beholder's presence in front of the image. As Fried puts it, the painter's task was to convince the beholder "that the beholder did not exist."[85] To be more precise, Diderot did not even always demand that the self should be forgotten; it was enough that the beholder forget the position his self was in. The latter is most evident in Diderot's descriptions of the landscapes by Claude-Joseph Vernet, which Diderot imagined himself to be inside.[86]

However, the artist was now suddenly called upon to depict his figures in a state of absorption. This was not to achieve some pictorial effect to help the beholder make sense of the drama inherent in the painting; on the contrary, the intention was to defeat the theatricality of the art of painting itself. Perhaps for the first time, someone had demanded that artworks should be made formally to fit with the problem posited by the beholder's consciousness of self. At least, the question's open ideology was new.[87]

The value of this development for us lies in the fact that Diderot's discourse, while presenting formal demands for a work of art, condensed around the same questions as when Diderot became concerned about the human-animal border and visuality in general. Mostly, these connections were of course left implicit, but one should not overlook the idea that Diderot considered the particular gaze he was after to be more "natural." As Fried can be seen to have hinted, such idea of naturality was, for the time being, a characteristic of French discourse; it may not be confused with the ideas prominent in Germany with which the naturality of self-expression and its suitable conditions were praised. The naturality, which Diderot was after, had much more to do with the task of destroying the theatrical milieu of life for the favor of a "*naïve imagination*" that could form a distinct kind of "drama" that Diderot favored for its *naturality*.[88] "*Without naïveté*," he wrote, "*there can be no real beauty*."[89]

It is, then, with the naïve gaze that Diderot attacked the mannerist aesthetics of his time. *Naïveté* was a condition which the monist naturalist came through. According to Diderot, there were no difference between being "a naïve hero, a naïve villain, naïvely religious, naïvely beautiful, a naïve orator or a naïve philosopher,"[90] nor being "naïvely . . . a tree, a flower, a plant or an animal."[91] This was a huge step, since Diderot's idea of naïveté, a natural image, is not based on the optical machine: "Naïvety means a great resemblance between the imitation and the thing, along with a great fluency of style . . . it means water taken from the stream and splashed straight onto the canvas."[92] Thus, style was heralded but not linguistics, since the naïve gaze abandoned the signifying process supporting the allegorical structure. In the same pages that we have just quoted, Diderot first attacked Boucher's paintings for their hypocritical nature but ended up praising his images of children, which nevertheless "are very naïvely childlike."[93] The machinery of imaginary identification, or of losing oneself with one's own gaze, now determined value.

Finally, one must see that if Diderot's extreme monism even contested the boundaries between animate and inanimate nature, the human-animal boundary was still invoked whenever Diderot needed to underpin his arguments with it. The naïve condition, already present in Diderot's theory of the hieroglyph, was driven by a need for natural visual of communication. Although the animal was not a major theme in Diderot's texts on art, its concept became for the first time entangled with the formalist demands of a theory that put its emphasis on a certain relation essentially favorable to the visual artistic experience. Diderot tried, in one of his texts, to maintain this minor difference between man and animal in connection with the experience of art. "There can be taste without sensitivity," he wrote,

> just as there can be sensitivity without taste. In its extreme form sensibility is not discerning; it is moved by everything. The first will tell you coldly, "That is beautiful"; the other will be moved, transported, intoxicated. "Saliet tundet pede terram, ex oculis stillabit amicis rorem." He shall stammer; he shall be unable to find expressions to describe the state of his soul . . . The happier of the two is undoubtedly the latter. The better to judge? That's another matter. Cold and austere men who tranquilly observe nature are often better able to articulate their views. They simultaneously are and are not enthusiasts; they are both, man and animal.[94]

The visual nature of animal apprehension towards the world was here determined as being an other compared to cold human self-consciousness. If there, in the context of Diderot's politics, was to be found a natural resistance towards the theatrical, this naïve gaze could be just as easily manifested as an animal way of looking onto the world. Of course, some dialectic was still needed for an animal to acquire a place that, ideally, could offer the possibility of living naturally within our own human culture. For Diderot, that place was provided by articulation and self-consistent rhetoric: "the capacity to be easily strongly moved by works of art represents a response at the animal level," as David Funt observes.[95]

A Concluding Note

Diderot was of course a master of essayist art criticism, a genre that has given its characteristics to the whole history of what we call the French theory of art. The essayist form is not irrelevant; ever since Montaigne's *Apology*, essayist writers had had freedom to examine questions concerning the human-animal border other than those suited to more formal contexts, and this should of course be kept in mind. My argument is nonetheless that the theories now introduced were paving the way towards the use of the concept of the animal that the high modernist avant-garde would soon adopt, and even they, in their historical context, show its logical relation to revolutionary politics, which in the field of the arts was often made to coincide with romanticized pictures of the domain outside hierarchical uses of language.

Perhaps more important, for the time being, was the new gaze that was imposed on the visual arts. At least after Étienne La Font de Saint-Yenne's groundbreaking *Réflexions sur quelques causes de l'état présent de la peinture en France* (1747) the French discourse on the arts had turned towards the questions that implicated the beholder of the image in a new manner. The symptoms were already visible in the works of the early 18th-century Academicians, but they were released into the open only when art criticism became an integral part of the public discourse.[96]

"The radical move that distinguishes the art of French modernity from what had gone before,"[97] as Fried has called it, was emblematized by these turns. The initial move was consistent with Diderot's "struggle against theatricality" that asserted that arts should "suspend or neutralize or undo the very condition of [theatrical, or self-conscious] spectatordom."[98] Therefore, it was not sufficient anymore to base theories of art on a theory of representation, but to formulate questions concerning the visual on its own terms, as prescribed by the relation of the beholder to the given artwork.

These ideas also evidently led to a situation in which there arose an opposition between the symbolic (the allegoric) and the visual register, though the opposition was still not presented in the definitive version that it took during high modernism.[99] Quite paradoxically, this condition also proved to become a cornerstone of the idea of the interconnectedness of all the arts. Although the beholder's gaze was never cut off from the intellectual skills to which it was attached—both Rousseau and Diderot more or less demanded a dialectics between the two—they understood the arts through the visual relationship the critic had made between himself and the work. The condition was, a bit surprisingly, slightly animalized. Rousseau and Diderot even made some efforts to understand animality through the predominantly visual identification with the natural world—evidently, regardless of the animal species. Such was perhaps the way pointed out by Condillac's theory of the animal, which Rousseau and Diderot honored.

Anyhow, the connections were mostly purely theoretical ones, as they were a symptom of a collision of different discourses.[100] The conceptual developments didn't have much influence on the work of the *animaliers* for the time being. Still, it is good to remember that animal art carried within itself a possibility of resistance towards the old hierarchies dictated by the Académie, which determined the hierarchy according to which animal art was supposed to be marginal: it was the lowest category in the canon led by history painting, a condition that was most severely dominant in France.[101] This also meant that there was always a resistance emerging when the genres and their rules were suppressed. One notable example of this sort of contestation can be found in the paintings of Jean-Baptiste Oudry: while illustrating Jean de La Fontaine's *Fables*, he made also animal paintings that were, in their tone and scale, made through formal means usually reserved for history painting.[102] Perhaps more assertively, the gesture was used nearly a century later in Théodore Géricault's post-Revolutionary paintings *The Wild Horse Race at Rome* and *Riderless Racers at Rome*, in which the genre of history painting was used pointedly to depict a scene of animal abuse.[103]

However, if Géricault, as has been suggested,[104] related himself to the pro-animal topics of his day, he might also have contributed to the topic that we have been examining here. What I would like to suggest, as a conclusion to this chapter, is that in Géricault's animal paintings there is already some sort of an avant-garde

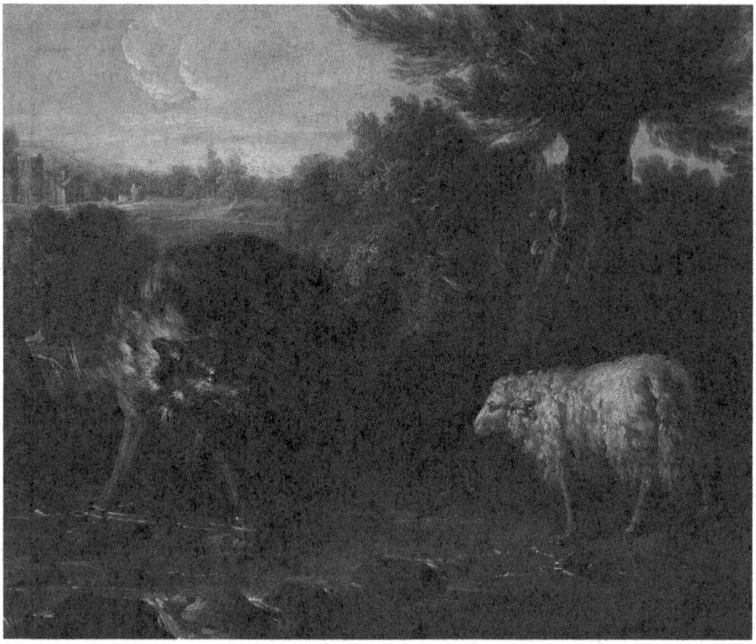

Figure 2.3 Jean-Baptiste Oudry, *The Wolf and the Lamb*. c. 1750. Oil on canvas, 104.1 x 125.7 cm. Private collection, Limoges. Available from: https://commons.wikimedia.org/wiki/File:Jean-Baptiste_Oudry_-_Le_loup_et_l%27agneau.jpg?uselang=fi (accessed November 7, 2016).

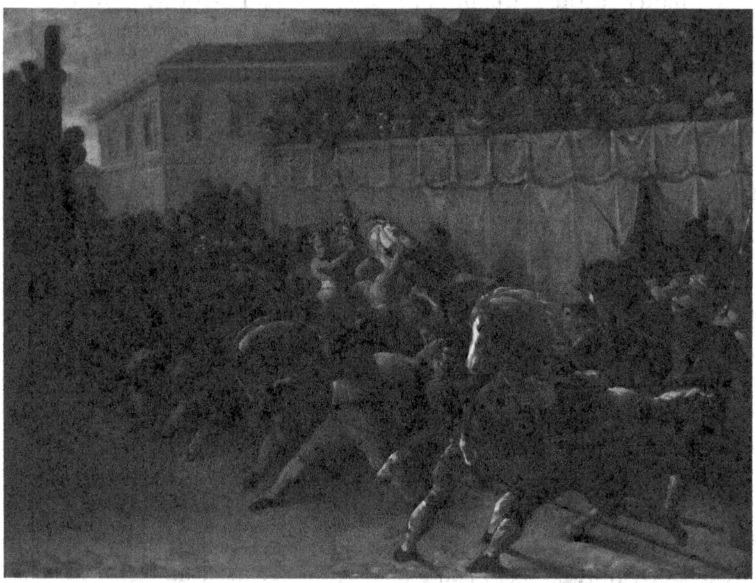

Figure 2.4 Théodore Géricault, *Riderless Racers at Rome*. 1817. Oil on paper, mounted on canvas, 44.9 x 49.5cm. The Walters Art Museum, Baltimore, MD. Available from: https://commons.wikimedia.org/wiki/File:Théodore_Géricault_-_Riderless_Racers_at_Rome_-_Walters_37189.jpg?uselang=fi (accessed December 1, 2016).

confusion at play facilitated by the discourse brought forward in Rousseau's and Diderot's assertions. As Fried has already remarked, the fact shines through from the fierceness of the wild beasts and military horses in Géricault's paintings—a quality that immediately elicits the question of animal otherness. In Géricault's Romanticist mind, according to Fried, the animal fierceness stands as "relate[d] intimately to human impulses and desires, far surpass[ing] human capabilities,"[105] but at the same time it can, as a consequence of the otherness depicted, "escape being perceived as theatrical, as grimace."[106]

If Fried's account has truth in it, as I think it might have, it shows a piece of the missing link between the accounts of the animal in high modernity and Diderot's and Rousseau's politics. Of Géricault's extremist interpretations of the beautiful but violent powers of fierce animality, always on the verge of a revolution, one could finally work out a scene in which the animal has become a model for the modern art to come. The animal living a life outside human rationality, openly visual, pure otherness from the start, following Fried, comes to take a role in Géricault's "tragic pictorial universe: [it] becomes an ideal of humanness that ultimately lies beyond human reach."[107]

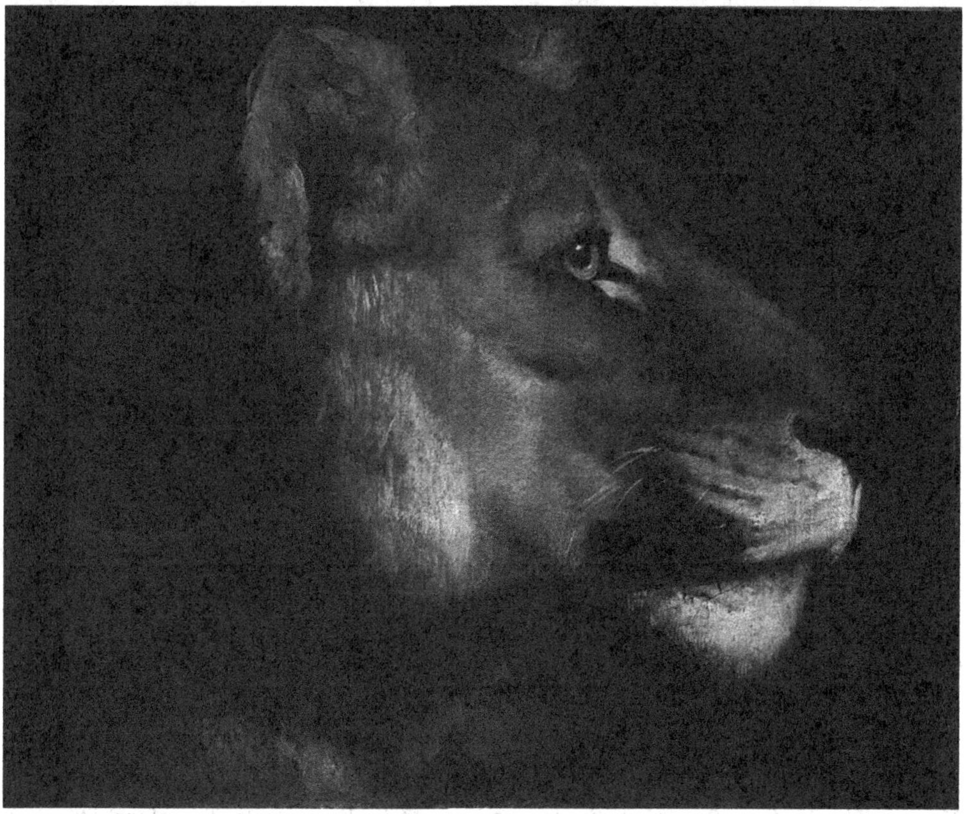

Figure 2.5 Théodore Géricault, *Head of a Lioness*. Early 19th century. Oil on canvas, 55 x 65 cm. Musée du Louvre, Paris. Available from: ARTstor, www.artstor.org (accessed November 7, 2016).

Notes

1. Jean-Jacques Rousseau, *The Social Contract & Discourses*, ed. G.D.H. Cole (London: J.M. Dent, 1920), 148.
2. Ibid., 144–145.
3. It is rarely noticed how German aesthetics had little influence in France for a whole century. Indeed, "aesthetics" was not mentioned in the *Dictionnaire des sciences philosophiques* until 1835, and nine years later the "*Dictionnaire*" added that "even if [aesthetics] has been cultivated with ardour in Germany for the last fifty years, its name has only lately started to get some attention [in France]." (For a discussion on the topic, see Jacqueline Lichtenstein, *Les raisons de l'art. Essais sur les limites de l'esthétique* [Paris: Gallimard, 2014]; for the citation, see Adolphe Frank, ed. *Dictionnaire des sciences philosophiques*, vol. 2. [Paris: Hachette, 1844–1852], 293.)
4. For an analysis of the scopic nature of the court in its connections with the attitude towards the visual that sprang forth in French theory, see Martin Jay, *Downcast Eyes: The Denigration of Vision in Twentieth-Century French Thought* (Berkeley: University of California Press, 1993), 83–148.
5. Ibid.
6. See ibid., 46–50.
7. With "disposition" I mean the structural setting which made it possible that the two separated fields could be interpreted against each other.
8. Norman Bryson, *Word and Image: French Painting of the Ancien Régime* (Cambridge: Cambridge University Press 1981), 1–28.
9. Ibid.
10. Ibid., 122. Bryson's analysis follows Ferdinand de Saussure's division between the two levels of a "sign": "signifier" and "signified."
11. Ibid., 87–88.
12. Ibid., 87.
13. Ibid., 88. This turn had also its theoretical equivalents inside the *Académie*. At least the influence of the empiricist turn should be mentioned, since it emphasized the relation of the beholder to the work of art at the expense of the means to produce certain significative patterns, gaining ground for modern aesthetics. Roger de Piles is a prominent example of this current. On de Piles's place in this sense in the history of the theory of art, see e.g. Lichtenstein, *Les raisons de l'art*.
14. Bryson, *Word and Image*, 88.
15. Denis Diderot, *Salons*, ed. Michel Delon (Paris: Gallimard 2008), 469.
16. Ibid., 122.
17. Johann Gottfried von Herder, "Essay on the Origin of Language," in *On the Origin of Language*, trans. John H. Moran & Alexander Gode (Chicago: University of Chicago Press, 1986), 103.
18. On Condillac's influence on Rousseau, see François Dagognet, *L'animal selon Condillac: Étude sur le "Traité des animaux"* (Paris: J. Vrin, 2004), 84.
19. The expression comes from the Comte de Buffon, to whom Condillac addressed much of his critique in *Traité des animaux*. On Condillac's counter-argument, see Étienne Bonnot de Condillac, *Traité des animaux* (Paris: J. Vrin, 2004), 113–141 and Dagognet, *L'animal selon Condillac*, 67.
20. Though it is well-known that Condillac's arguments on animal nature vary according to their context, and that they develop greatly between *Essai sur l'origine des connaissances humaines* (1746) and *Traité des animaux* (1755)—I am treating them here as a constitutive whole to give an account of the possibilities that his claims on animal nature were able to open for the theoreticians to come. For a larger account of Condillac's views on the animal in context, see e.g. Dagognet, *L'animal selon Condillac*.
21. Condillac, in Gary Steiner, *Anthropocentrism and Its Discontents: The Moral Status of Animals in the History of Western Philosophy* (Pittsburgh, PA: University of Pittsburgh Press, 2005), 177; originally in Étienne Bonnot de Condillac, *Essay on the Origin of Human Knowledge*, trans. Hans Aarsleff (Cambridge: Cambridge University Press, 2001), 39.

22. Ibid.
23. Condillac, *Traité des animaux*, 200.
24. Cited in Steiner, *Anthropocentrism and Its Discontents*, 176; originally, in Condillac, *Essay on the Origin of Human Knowledge*, 37.
25. Ibid.
26. Condillac, *Traité des animaux*, 200.
27. Ibid.
28. See e.g. Condillac, *Essay on the Origin of Human Knowledge*, 41.
29. Condillac, *Traité des animaux*, 200.
30. Ibid., 132.
31. Ibid., 130–131.
32. For Rousseau's somewhat paradoxical attitude towards animals, see e.g. Jean-Luc Guichet, *Enjeux de la question de l'animal sous les Lumières: Condillac, Diderot, Rousseau*, 6–9, accessed January 10, 2015, http://ecole-thema.ens-lyon.fr/IMG/pdf/Article_Guichet-2.pdf.
33. On Condillac's influence on Rousseau, see Dagognet, *L'animal selon Condillac*, 84.
34. Jean-Jacques Rousseau, *On the Origin of Inequality*, trans. G.D.H. Cole (New York: Cosimo, 2005), 34. Rousseau writes: "Every animal has ideas, since it has senses; it even combines those ideas in a certain degree; and it is only in degree that man differs, in this respect, from the brute. Some philosophers have even maintained that there is a greater difference between one man and another than between some men and some beasts." Although Rousseau's reasoning, as in the following citation, is based on the exclusively human capacity to act as a free agent, it is notable that he has decided to refer to Plutarch's famous pro-animal phrase that invokes the greater difference between one man and another than between a man and an animal, which was also used as a departure by Montaigne.
35. Ibid., 34.
36. Ibid., 35.
37. Ibid., 36–37.
38. Ibid., 62.
39. "There is another very specific quality, which distinguishes man and animal, and which will admit of no dispute. This is the faculty of self-improvement [se perfectionner]." (Ibid., 35).
40. Condillac also maintained that animals, unlike humans, are not able to correct their habits by themselves. Condillac, *Traité des animaux*, 188–195.
41. Rousseau, *On the Origin of Inequality*, 44. My emphasis.
42. For a commentary on the opposition between language and the state of pure nature in Rousseau's thinking, see Jacques Derrida, *Of Grammatology*, trans. Gayatri Chakravorty Spivak (London: Johns Hopkins University Press, 1976), 253.
43. Jean-Jacques Rousseau, *Émile, or On Education*, trans. Barbara Foxley (Sioux Falls, SD: NuVision Publication 2007), 81.
44. I am referring to Rousseau's *Essay on the Origin of Languages*. The *Essay* seems to have been written mostly in 1754, though it was not published until 1781. The tensions between the work and the *Discourse on Inequality* have often been noted—though the *Essay on the Origin of Languages* was first meant to form part of the *Discourse on Inequality*, Rousseau thought it to be "too long and out of place" (Jean-Jacques Rousseau, *"The Discourses" and Other Early Political Writings*, ed. Victor Gourevitch [Cambridge: Cambridge University Press, 1997], 393)—and Rousseau himself is known to have revised the work until 1763, when it was sent to its publisher, one year after *Émile's* publication. One should also remember that Condillac's *Traité des animaux* was published in 1755.
45. Jean-Jacques Rousseau, "Essay on the Origin of Languages," in *On the Origin of Language*, trans. John H. Moran & Alexander Gode (Chicago: University of Chicago Press, 1986), 5.
46. Ibid., 32.
47. Rousseau, *Émile*, 39.
48. Ibid., 49.
49. Ibid.
50. Rousseau, "Essay on the Origin of Languages," 32.
51. Ibid., 12.
52. Ibid.

53. Ibid., 12–13.
54. Ibid., 13.
55. Ibid., 12.
56. Rousseau, *The Social Contract & Discourses*, 191.
57. Rousseau, "Essay on the Origin of Languages," 12–13.
58. Love "might also," as Rousseau reminds us, have "invented speech," but it did not, and luckily so, since the development would have then taken "a less happy course." (Ibid., 6.)
59. Ibid.
60. Ibid. Translation modified.
61. See ibid., 10. This curious remark is affirmed by Rousseau, making fun of the idea that the differences between animal and human speech might be explained by the organic differences between the species: "It is said to be explicable by organic differences. I would be curious to witness this explanation." (Ibid.)
62. Ibid.
63. Ibid., 55. "But what makes painting an imitative art? Drawing. What makes music another? Melody."
64. Ibid.
65. See ibid., 63.
66. Ibid.
67. Ibid. My emphasis.
68. Rousseau, *Émile*, 77.

It should be noted that Jacques Derrida, to give a famous example of a possible other interpretation of Rousseau's remarks on the human-animal boundary, understands that Rousseau claims that animality has nothing to do with imagination and *mimesis*. Derrida even maintains that the two even make up the difference between animals and humans in Rousseau's mind. He interprets Rousseau's constant remarks about imagination and *mimesis* being the foundation of human speech—in other words, the origin of language—as signifying that animals are only capable of tasks that are dependent on reasoning and self-interest, while men can speak by going beyond their own interests because of their empathically oriented imagination. In Derrida's words: "Animals, whose pity is not awakened by the imagination, have no affinity with the other as such. That is why there is no animal music. One speaks of animal music only by looseness of vocabulary and by anthropomorphic projection. The difference between the glance and the voice is the difference between animality and humanity. Transgressing space, mastering the outside, placing souls in communication, voice transcends natural animality" (Derrida, *Of Grammatology*, 195–196). As I have hoped to show, this is not usually the case with Rousseau. On the contrary, Rousseau keeps insisting, at least after the publication of the *Discourse on Inequality*, on his idea about how mimetic activity and visual language play a central role in animal life.

It is nonetheless true that the important place that the concept of the animal has in Rousseau's views about the origin of languages is also seen by Derrida. Following him, Rousseau has an idea of the supplementary nature of human cognition based on (human) imagination that is by its turn the foundation of all mimetic activity. Derrida criticizes Rousseau above all for the fact that Rousseau did not let go of the classical metaphysical disposition that kept terms like exteriority and interiority, life and death, speech and writing, etc. apart from each other, while Rousseau nevertheless kept a "dream that consisted of making the supplement enter metaphysics by force" (Ibid., 315). What Derrida doesn't notice, while insisting on his thesis of the exclusively human ability to imagine, is that the visual language that Rousseau grants to animals erects itself as a space of communication inside human language, opening human communication to non-conventionalized significations that are nonetheless mimetic. Derrida almost says this when he analyzes the nature of the supplement in Rousseau's text: "That confirms what we proposed above: imitation cannot allow itself to be appreciated by a single act. Rousseau has need of imitation, he advances it as the possibility of song and the emergence out of animality, but he exalts it only as a reproduction adding itself to the represented though it *adds nothing*, simply supplements it. In that sense he praises art or *mimesis* as a supplement. But by the same token praise may instantly turn to criticism" (Ibid., 196). Derrida only fails to see that if the supplementary nature of *mimesis* "adds nothing," it is only because Rousseau puts his hopes in a mimetic visual language that would not surrender to the vicissitudes of historical human developments.

69. It may still be noted that Rousseau was not the only one in his time to claim that the origin of language resided in the visual register. Giambattista Vico, for example, had already tried to prove it in the first decades of the century. Vico had thought that imagination ["fantasia"] was the quality that stood as the marker of the exclusively human sphere. Relatedly, Derrida humorously explains the etymology of the idea and its appearance in Rousseau's texts: "Although this proposition [that the visual was understood as the origin of language] was not peculiar to Rousseau, although he might have encountered it in Vico, although he must not only but surely have read it in Condillac who must not only but surely have taken it from Warburton, we must emphasize the originality of the *Essay*" (Ibid., 272). This originality is, in my view, in Rousseau's capability of inserting Condillac's ideas to the developments at hand.
70. Diderot cited in Bryson, *Word and Image*, 1981, 180; originally, in Denis Diderot, *Salons*, vol. 3, ed. Jean Seznec & Jean Adhémar (London: Clarendon Press 1979), 190.
71. Diderot Studies XI, in David Funt, "Diderot and the Esthetics of Enlightenment," in *Diderot Studies XI*, ed. Otis Fellows & Diana Guiragossian (Genève: Librairie Droz, 1968), 171; originally in Diderot, *Salons*, vol. 3, 157.
72. Denis Diderot, *A Conversation between D'Alembert and Diderot*, trans. Ian Johnston (Nanaimo, BC: Johnstonia), accessed January 2, 2015, www.oneeyedman.net/school-archive/classes/fulltext/www.mala.bc.ca/~johnstoi/diderot/conversation.htm.
73. "And you speak of essences, you poor philosophers! Forget about your essences," Diderot writes in *Rêve d'Alembert* (Denis Diderot, *Rêve d'Alembert*, trans. Ian Johnston [Nanaimo, BC: Johnstonia], accessed January 2, 2015, http://records.viu.ca/~johnstoi/diderot/revedalembert_tofc.htm). Later on, he affirms his views on the subject by allowing an ability to feel for even for the minutest particles in the living world, doing this in a self-evident anti-Cartesian gesture: "[—] Each form has the happiness and unhappiness appropriate to it. From the elephant all the way to the aphid . . . from the aphid all the way to the sensitive and living molecule, the origin of everything, there's no point in all nature which does not undergo pain or pleasure" (Ibid., 15).
74. Diderot, *A Conversation between D'Alembert and Diderot*. The idea of an animal that can retain the "memory of its actions" follows the guidelines of Diderot's argument, formed in contrast to the ones given by Descartes, who had considered animal life in terms of passivity. According to Diderot, "the statue only has latent sensation, and man, animals, and perhaps even plants are endowed with an active sensibility" (Ibid.).
75. Of Diderot's concept of fibre, see Tobias Cheung, "Omnis Fibra ex Fibra: Fibre Economies in Bonnet's and Diderot's Models of Organic Order," in *Transitions and Borders Between Animals, Humans and Machines 1600–1800*, ed. Tobias Cheung (Leiden: Brill, 2010), 66–70.
76. Diderot, *Rêve d'Alembert*.
77. See e.g. Bryson, *Word and Image*, 176.
78. Diderot, *A Conversation between D'Alembert and Diderot*.
79. See Kenneth Berri, "Diderot's Hieroglyphs," in *SubStance*, *Issue 92* (vol. 29, no. 2), 2000: 73–74.
80. Diderot cited in ibid., 73.
81. See e.g. ibid., 72–75.
82. Bryson, *Word and Image*, 193.
83. Ibid., 122. Likely in the same way, Diderot himself established in his dialogues and in his experimental novels, the twofold strategies with which to attack the self-conscious subject claiming to total self-consciousness—one need only to think of Diderot's novel *Jacques le fataliste*. For an interesting article on the subject of Diderot's conjectural experiments and strategies, see François Pépin, "Diderot et la conjecture expérimentale, ou comment concilier l'extravagance de l'enthousiaste et la prudence de l'authentique génie," in *MLN*, vol. 129 no. 4 (September 2014).
84. Diderot, in Bryson, *Word and Image*, 202; originally in Denis Diderot, *Salons*, vol. 4, ed. Jean Seznac & Jean Adhémar (London: Clarendon Press 1979), 88.

85. Michael Fried, *The Moment of Caravaggio* (Princeton, NJ: Princeton University Press, 2010), 121; on the topic at large, see Michael Fried, *Absorption and Theatricality: Painting and Beholder in the Age of Diderot* (Chicago: Chicago University Press, 1980), 121.
86. See for instance Diderot 2008, 272–376. Fried writes quite descriptively: "In Diderot's accounts of certain pictures that were unsuited to dramatic rendering because of their subject matter, he took up the fiction that he himself was literally inside the painting . . . This suggests that for Diderot the aim of such works was not to exclude the beholder . . . but on the contrary to draw the beholder into the representation—a feat that would equally have denied his presence before the canvas (and of course, again paradoxically, would equally have transfixed him precisely there)." Michael Fried, *Courbet's Realism* (Chicago: The University of Chicago Press, 1990), 8.
87. Fried writes: "My point, however, is that starting around the mid-1750s in France (and only there) the inescapableness of beholding, or say the primordial convention that paintings are made to be beheld, became deeply problematic for the enterprise of painting precisely to the extent that the latter took upon itself the task of striving to defeat what Diderot called theater, and that the irruption of that internal conflict or contradiction was something new in the history of art" (Fried 1990, 12–13). The problems only started when this characteristic allowed itself to be read as a sign of transparent self-consciousness: "In other words, absorption became a critical value just when it could no longer be taken for granted as a pictorial effect. And it could no longer be taken for granted as a pictorial effect because for the first time it was called upon to defeat theatricality." (Ibid.)
88. See e.g. Denis Diderot, *Salons*, ed. Michel Delon (Paris: Gallimard, 2008), 468–470.
89. Ibid., 469.
90. Ibid.
91. Ibid.
92. Ibid.
93. Ibid.
94. Ibid., 234.
95. Funt, "Diderot and the Esthetics of Enlightenment," 60.
96. Étienne la Font de Saint-Yenne, "Réflexions sur quelques causes de l'état présent de la peinture en France, avec un examen de principaux ouvrages exposés au Louvre," in *Œuvres critiques*, ed. Étienne Jollet (Paris: ENSBA, 2001). See e.g. Lichtenstein, *Les raisons de l'art.*
97. Fried, *Courbet's Realism*, 178–179n.
98. Ibid.
99. Later, this would lead to an almost total separation between the concepts of the allegory and the symbol. We will deal with the question in connection with our topic in Part 3.
100. One should note that Diderot's art criticisms—"The Salons"—were available only to a small group of readers before the late publication of the texts in 1798.
101. Evidently, this did not mean that there were not animal lovers among some of the most lauded artists of the time, and some of the artists were not able to resist the temptations of the Cartesian frame of thought with regard to animals. See for example Sarah R. Cohen, "Animal Performance in Oudry's Illustrations to the Fables of La Fontaine," in *Studies in Eighteenth Century Culture,* vol. 39, 35–76.
102. See e.g. ibid., 36.
103. See e.g. Stephen F. Eisenman, *The Cry of Nature: Art and the Making of Animal Rights* (London: Reaktion Books, 2013), 156.
104. See e.g. ibid., 155–161. Eisenman lists Géricault's works that depict scenes of horse abuse. Among them are: "Horses Going to a Fair" from *Various Subjects Drawn from Life and on Stone* (1821), "Horses Being Led to the Slaughterhouse" from *Suite of Eight Small Works* (1823), *Dead Horse* (1823), *Race of the Riderless Horses in Rome* (1817), *Butchers of Rome* (1817), *The Plasterer's Horse* (1822).
105. Fried, *Courbet's Realism*, 24.
106. Ibid.
107. Ibid.

3 Art and Evolution

> Metaphysics must flourish—He who understand [sic] baboon would do more toward metaphysics than Locke.[1]
>
> —Charles Darwin, *Notebooks*

Introduction: Darwin's Century

While the crisis of the allegory went on—in our case, mainly, the crisis of animal symbolism—art also had to cope with two scientific trends that often went hand in hand: materialism and Darwinist evolutionism. As the materialist ideas largely framed the discourse of avant-garde art that started to attack Romanticist sensibility and metaphysical idealism from the mid-century, Darwinist ideas, as we shall see, contributed to the discourse by calling into question the singularity of human art and the uniqueness of man's experience of beauty. Perhaps a little bit surprisingly, the Darwinist ideas found themselves, at the end of the century, at least in some circles, in a central position regarding the general thoughts of art's essence. In one sense, this was a natural continuation from the fact that it had become a commonplace to discuss the practice of art in relation to the human-animal border, though mostly it was done in negative terms. This section might be considered an account of these tendencies.

The picture is of course not so simple as that: the Darwinist and materialist philosophies did not just throw out the old idealist habits. Rather, their mutual emergence launched a minor tradition in the philosophy of the arts, a tradition that took an important place among the artistic revolutions at the turn of the 20th century. Mostly its influence was felt when the new frames started to be integrated into the major philosophical systems that dominated the discipline of aesthetics and the theoretical discourse on the arts—namely, Romanticism, positivism, and phenomenology.

Evidently, the three had already contributed to the newly formed understanding of the relation between man and nature. As Margot Norris has noted, positivism (in its anti-idealism), Romanticism (in its anti-rationalism), and phenomenology (in its anti-metaphysicality) had attached themselves to new quests that tried to get rid of worn-out views, and they were now asking questions about worldly immanence.[2] The problem, in contrast to the concept of the animal, lay in the nature of the immanence they all were eager to touch upon; as we have partly been examining in the first part of this work, they shared a continuing analysis of man's immanent relation with nature, but by leaving animal consciousness to a state of pure otherness by holding on to the old differences between man and animal that were thought to be essential—the abilities of consciousness, etc.

While discussing the history of animals and visual culture in the 19th century, one should not forget an essential question, though its theoretical contribution may seem insignificant at first sight: the question of animal rights and of pro-animal sentiments, which was also discussed during the era. Although the question is not central to our topic—most of the contemporary criticism of modern attitudes attaches itself to the fact that questions on animal rights did not contribute to the field of the arts in the modern era—one should be reminded that the idea of the legal rights of animals did change our view of them. This challenge was nowhere so visible as in England in the 19th century. Though it had long been commonplace to argue—as, for example, Kant had done—that it was man's obligation towards himself to be kind towards animals,[3] the question was understood in a different light when the sentience of the animals were brought up anew. Most notably this was done by Jeremy Bentham, who argued that animals were "sentient beings" like humans. "The question is not, Can they reason?, nor Can they talk? but, Can they suffer?"[4] Bentham wrote in a famous footnote in his *Introduction to the Principles of Morals and Legislation* in 1780.[5]

It was also in England where questions about cruelty towards animals were most substantially raised in art, long before the establishment of the Royal Society for the Prevention of Cruelty to Animals (1824) or the first laws restricting man's cruelty towards the beasts ("Cruelty to Animals" Act, 1849). William Hogarth's series of engravings called *The Four Stages of Cruelty* (1751), William Blake's illustrations and poems in *Songs of Innocence and of Experience* (1789), and Thomas Bewick's pictures depicting maltreated horses are perhaps the most well-known examples.

Of course, the mood was quite selective, as it always is, and it favored the species to which the English social life was connected, i.e., dogs and horses. Sometimes, the two had their moments in the art of portraiture, even without a reference to their human owners (a distinguished example is George Stubbs's 1762 portrait of the Marquess of Rockingham's racehorse Whistlejacket). This tendency also had its consequences, even leading animal art to be devalued once again. The fashion for sentimentalizing animal figures (notorious examples can be found in Edwin Landseer's and Rosa Bonheur's works) became degraded in 19th-century conceptions of good art, after the discrimination of sentimentality among the avant-garde had taken place. Thus, when Joséphin "Sâr" Péladan, in a Symbolist context, wrote his rules for the 1891 *Salon de la Rose + Croix*, the ninth rule was that all paintings with domestic animals and those relating to sport were immediately rejected, while paintings with wild beasts could make it through.[6] We shall later examine some of the reasons for this.

To get started, a couple of points still need to be made. The appropriation of the ideas leading to Darwin's theory of evolution had started long before his work. While the 18th-century materialisms were developing, the principles of modern biology were also being born when the taxonomies of exterior appearances were dismissed for the sake of an analysis of material interior worlds.[7] Evidently, the theories didn't have much to say on the visual arts or vice versa, or not yet. Their influence was, however, carried over to the practices of image-making nonetheless. An easy example can be found in the work of George Stubbs, who, at the end of his career, made a series of anatomical drawings for his book, *A Comparative Anatomical Exposition of the Structure of the Human Body with that of a Tiger and a Common Fowl* (1795–1806),[8] a work often

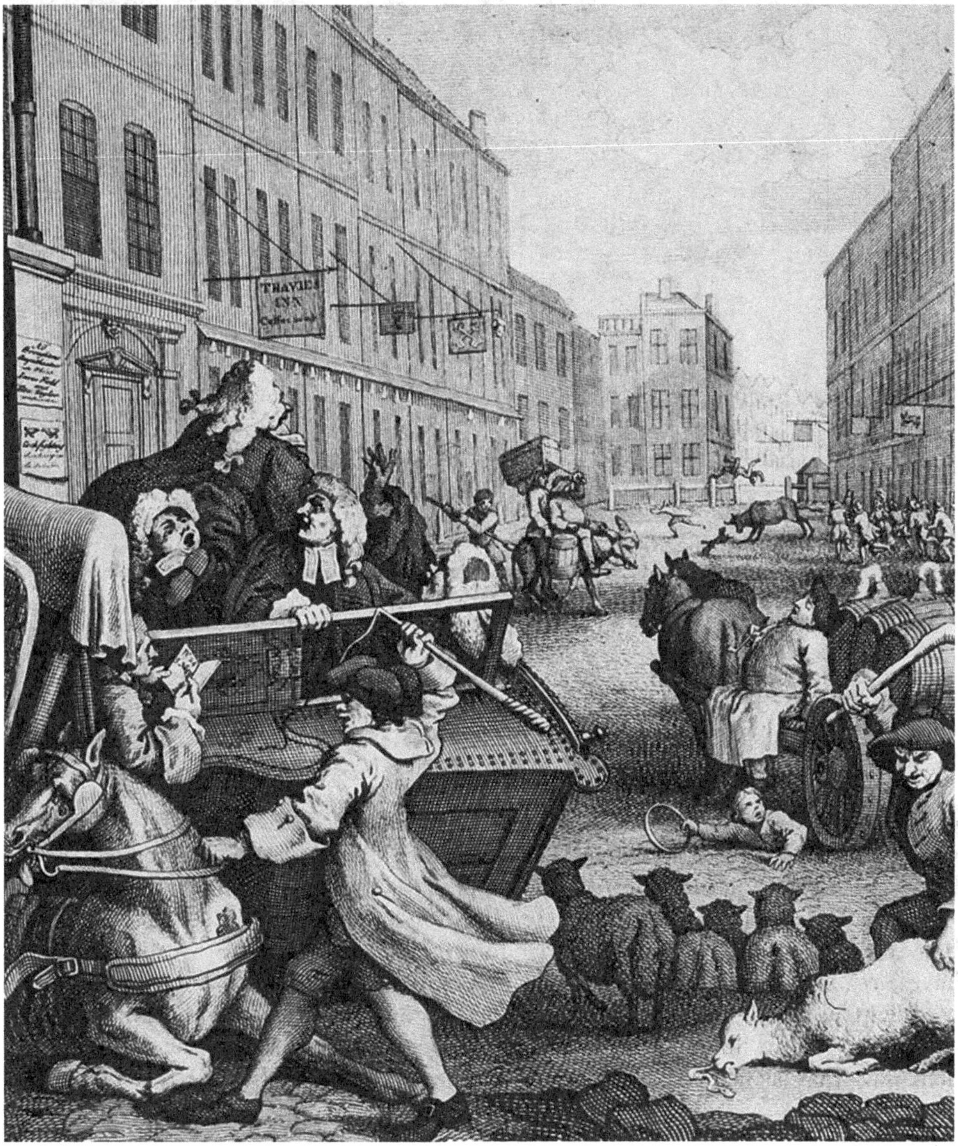

Figure 3.1 William Hogarth, *The Four Stages of Cruelty: Second Stage of Cruelty*. 1750–1751. Etching with engraving, 35 x 30 cm. British Museum, London. Available from: www.artstor.org (accessed December 1, 2016).

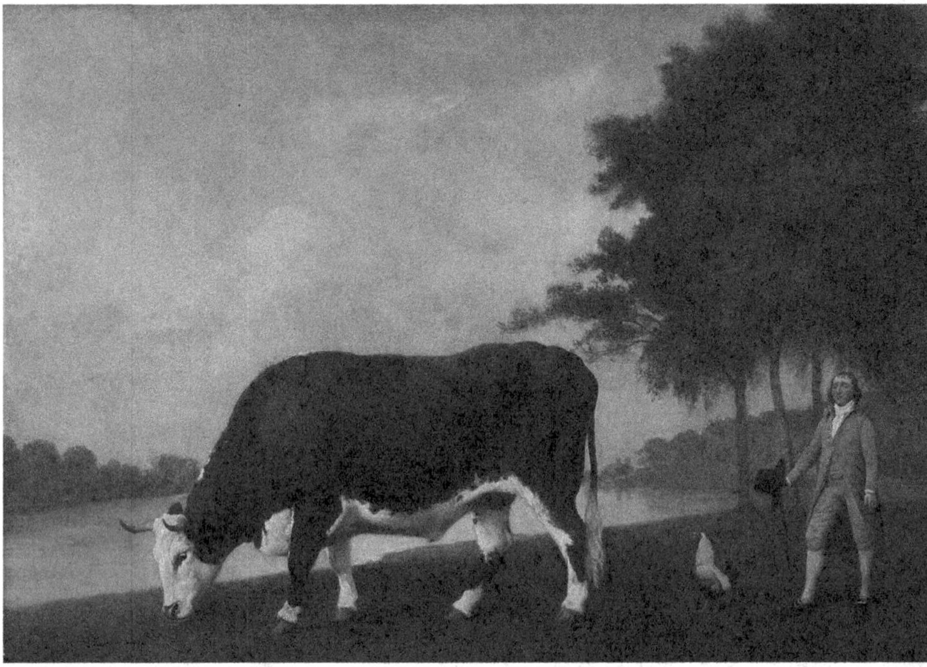

Figure 3.2 George Stubbs, *The Lincolnshire Ox*. 1790. Oil on panel, 67.9 x 99.0 cm. Walker Art Gallery, Liverpool. Available from: https://commons.wikimedia.org/wiki/File:George_Stubbs_-_The_Lincolnshire_Ox_-_Google_Art_Project.jpg?uselang=fi (accessed November 7, 2016).

paired with Stubbs's fantastically naïve late work, *The Lincolnshire Ox* (1790). The works can, as Stephen F. Eisenman has remarked, be considered as statements of the human life inhabiting the same ontological plane as the rest of animal creation.

The second point concerns the intensification of the question of the differences between animal and human scale that took place in the late 18th century. As we have already seen how the Kantian-Romanticist tradition established the question, we might now examine the resistance that it confronted. While still affirming the divide between the human scale and its other—namely, animal life—the opposing side demanded that beauty exceeding human measures should also be expressed in the arts.

The Subjective and the Objective

As Paul L. Sawyer has put it, the English mid-18th-century cultural discussions were epistemologically marked by the "intersection of Romantic tradition with the triumph of scientific naturalism".[9] From this collision, a new conception of the animal and the arts arose: a conception that Norris has even thought appropriate to call "biocentric."[10] The intersection of Romanticist subjectivism and scientific objectivism was emblematized by the work of Charles Darwin.[11] Darwin's theory of evolution became so infamous in its own time that not even the theorists of art—however vague their understanding of it—could disregard it. Soon, the distribution of Darwinian ideas, and the Victorian faith in the principles of scientific naturalism of which Darwin was part, started to dismantle the Romanticist frame, a frame according to which the world was to be looked at on the human scale. The process through which this confrontation came about was of course already taking place during Darwin's time, and to a large extent it surpassed the perspectives of his texts.

The oppositional frame with which the Romanticist apotheosis of human subjectivity became entangled, as Lorraine J. Daston and Peter Galison have noted, was partly a consequence of the birth of the ideal of objectivity among the scientific community.[12] The opposition, following them, culminated in Darwin's time, in England in the 1850s, when Romanticist thinkers had to counteract the tide.[13] In other words, the old interconnectedness between the subjective and the objective, which had colored the epistemic ideas of the Middle Ages by equating the subjective with the accidental, while representing objectivity as a necessary condition for the human subject to objectify the world, i.e., to understand it with universals—had now been overturned.[14] This reversal had taken place due to Kantian influences. In Kant's ideas about subjective judgments, subjectivism had not yet turned into an outright hostility against naturalist premises, a turn that ideas about subjectivity could only take after the rise of Romanticism in an age when the modern scientific ideal of objectivity had closed off the scientific subject from its territory.[15] It was not until the turn of the 19th century that the objective was identified with nature and the subjective with human intelligence.[16] Only after these developments could one, as Samuel Taylor Coleridge did, consider the two "conceptions . . . in necessary antithesis."[17] What was implicated in connection with the animal, as we have seen, was that it, in its naturality, was thrown into a passive objectivity differing from creative human subjectivity.[18]

In the artistic world, the resistance towards Romanticist views was expressed by an imagery demanding a different scale than the human one; to depict scenes from a state of things in which the human beholder did not, or did not have the power to, interfere. Along with the objectivist discourse, a demand to imagine a universe in which the human measures didn't count was felt throughout the different categories of thinking. Man was not—at least for some—the measure of things anymore. Perhaps this universe remained a fictional one, but it nevertheless contributed to new ideas in the arts, while claiming objectivity. This was the ground on which Norris could claim the biocentricity of the theories in question. The aesthetic quality, inherent in a world without humanity, had also been suggested by Darwin himself, when he had commented on the principle of beauty from the perspective of natural selection.[19] This paragraph could be considered as representing that paradigm:

> With respect to the belief that organic beings have been created for the delight of man,—a belief which it has been pronounced is subversive to my whole theory,—I may first remark that the sense of beauty obviously depends on the nature of the mind, irrespective of any real quality in the admired object; and that the idea what is beautiful, not innate or unalterable. We see this, for instance, in the men of different races admiring an entirely different standard of beauty in their women. If beautiful objects had been created solely for man's gratification, it ought to be shown that before man appeared, there was less beauty on the face of the earth than since he came on the stage. Were the beautiful volute and the cone shells of the Eocene epoch, and the gracefully sculptured ammonites of the Secondary period,

> created that man might afterwards admire them in his cabinet? Few objects are more beautiful than the minute siliceous cases of the diatomaceae: were these created that they might be examined and admired under the microscope? The beauty in this latter case, and in many others, is apparently due wholly to symmetry of growth. Flowers rank among the most beautiful productions of nature; but they have been rendered conspicuous in contrast with the green leaves, and in consequence at the same time beautiful, so that they might easily be observed by insects. Hence we may conclude that, if insects had not been developed on the face of the earth, our plants would not be decked with beautiful flowers; one kind open and coloured, destitute of nectar, and never visited by insects.[20]

Objectivist glances towards a world without a human subject, as manifold as they were, had their counter-reactions. From Samuel Taylor Coleridge and William Wordsworth to John Ruskin, the English Romanticist tradition resisted the scientific worldview in the name of culture and of the arts.[21] Coleridge's attack against the views implicated in Linnaean taxonomy work as an example: "Linnaeus, whose artificial system of classification was based on sexual characteristics of plants, failed to discover," Coleridge openly remarked, "the central idea of vegetation itself," with which he thought Linnaeus might have discovered the "constitutive nature and inner necessity of sex itself."[22] The conclusion was nothing but awaited, at least when compared retrospectively to Darwin's passage just cited: according to Coleridge, the moment "the full applicability of an abstract science ceases, the moment reality begins."[23] This kind of fantasmatic subjective reality may have proved to be the last resort of the Romanticist mind if the objectivist ideal hadn't been at least as fictionalized.

If accepted, Darwin's theory could of course have had violent consequences with regard to ideas of the human-animal boundary or the dualisms supporting it, but this was not often the case. Although Darwin had wanted to "attack" the old metaphysical "citadel," by concluding that "the mind is function of the body,"[24] Darwin's claims that the most exclusively human characteristics were also a consequence of biological evolution did not, unsurprisingly, resonate much in the theories of the arts.[25] Nonetheless, the popularity of the Darwinian ideas did result in the overdetermination of the question of the human-animal boundary.

If not immediately for the animal in particular, the objectivist deduction paved a way for the possibilities of viewing the natural in a different light. Among artists, these influences were felt the most strongly within the boundaries of landscape painting, as Sara Barnes and Andrew Patrizio have remarked, when the Romanticist idiom was confronted by new artwork whose imagery was full of the "monumentality of nature (mountainous ranges, deep river valleys) and singular natural events (the erupting volcano, the earthquake)."[26] This might also have equalled a revitalization of the aesthetics of the Burkean sublime that overthrew the human scale.

Edmund Burke, to consider his work on the sublime here somewhat anachronistically, had indeed left a remarkable heritage for those trying to glance past the restrictions imposed by human scales. Burke's early work, *A Philosophical Enquiry into the Origin of Our Ideas of the Sublime and Beautiful*, published in 1757, is an important document in this regard, since it stresses repeatedly that animal life has a central role in our understanding of the sublime in nature, and that the experience of the sublime is negatively attached to the questions of human proportions and scale. According to Burke's vision, animal life could be understood as sublime in itself because of its relative disproportion to human perspective. From giant scale to infinite minuteness,[27] animal life could, in Burke's view, give us a sense of the sublime:

> The passion caused by the great and sublime in nature, when those causes operate most powerfully, is Astonishment; and astonishment is that state of the soul, in which all its motions are suspended, with some degree of horror. In this case the mind is so entirely filled with its object, that it cannot entertain any other, nor by consequence reason on that object which employs it . . . There are many animals, who though far from being large, are yet capable of raising ideas of the sublime, because they are considered as objects of terror.[28]

What seems most remarkable in Burke's arguments is the margin that he leaves for the communication between men and animals, though one should first think of course of the preceding remark with relation to the Romanticist imagery of the wild beasts; "the tigers of wrath are wiser than the horses of instruction,"[29] as Blake had poetically put it, could be seen as an apparent implication of Burke's discourse. Behind it lies another idea, which presents animality as a sphere of otherness that, when confronted, could penetrate the human mind—evidently, because it affirms that animals act in a world beyond human understanding, beyond human linguistics. It was both the eye—the glance—that had, according to Burke, "so great a share in the beauty of animal creation,"[30] and the sound of an animal, or even a possibility of an appearance of an animal sound, that was perceived by Burke to be sublime because with them the animals could communicate with us on another level than with languages operating anthropocentrically. This is how Burke put it in the 1750s:

> *The Cries of Animals.* Such sounds as imitate the natural inarticulate voices of men, or any animals in pain or danger, are capable of conveying great ideas; unless it be the well known voice of some creature, on which we are used to look with contempt. The angry tones of wild beasts are equally capable of causing a great and aweful sensation . . . It might seem that these modulations of sound carry some connection with the nature of the things they represent, and are not merely arbitrary; because the natural cries of all animals, even of those animals with whom we have not been acquainted, never fail to make themselves sufficiently understood; this cannot be said of language. The modifications of sound, which may be productive of the sublime, are almost infinite.[31]

The ambiguity with which we face the things sublime to us, Burke concluded, shines through our relation with even finely domesticated animals like dogs, a truth proved by the degrading and paradoxical language with which we express our love towards them.[32]

To get back to the context in question, the Darwinist discourse indeed attached its ideas to the objectivist ideal of overcoming the boundaries of the human scale, setting itself in opposition with the Romanticist mission "to incorporate," in Wordsworth's terms, the passions of men "with the beautiful and permanent forms of nature."[33] As James Paradis has listed, the qualities of (1) human scale and orientation, (2) of permanence and harmony, and (3) of emotional and spiritual content, all heralded by the Romantics, were becoming displaced by the new ideologies introduced on objectivist or naturalist grounds.[34] Greatly affected by evolutionary ideas, these new trends were constructed upon views that valued (1) global scale, (2) impermanence, and (3) calculated physical organization in space and time.[35] A fine example of the current is William Dyce's *Pegwell Bay, Kent—a Recollection of October 5th 1858*, depicting a monumental landscape with Donati's comet in the sky, and named with scientific precision.

The divide between the two perspectives has been historically quite important, since it contributed to the developments through which Darwinian theory soon acquired its attractiveness among the radical progressivist tendencies in the art world. The questions discussed by Paradis were of course not the only ones that came about. Indeed, perhaps the most important of the consequences of these discussions was the formation of artistic Naturalism, which became deeply related to the crisis of symbolism. Frequently, some sort of objectivist "repression of metaphorical thinking"[36] did fit the demands of naturalist politics, although the naturalist views might often have operated on quite a non-scientific basis. However, to let nature speak, it was now necessary to get rid of the old formulas and of the old allegorical relation, since the new art had to overthrow familiar measures. If nature, in its objective form, "remain[ed] largely inaccessible to language,"[37] the objectivist naturalist view could shine through in an

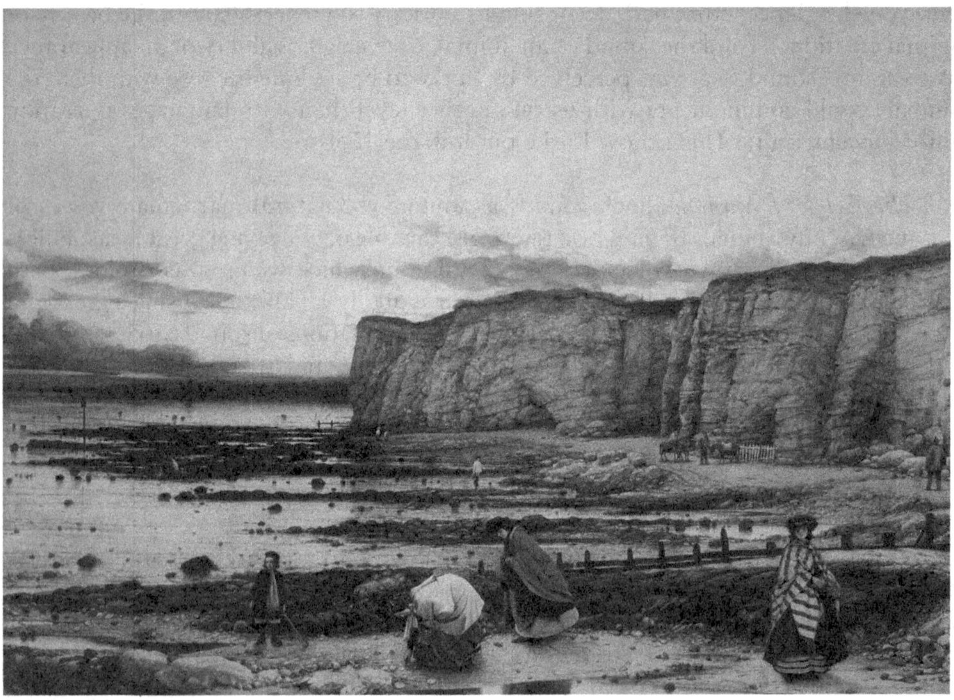

Figure 3.3 William Dyce, *Pegwell Bay, Kent—a Recollection of October 5th 1858*. c. 1858. Oil on canvas, 63.5 x 88.9 cm. Tate Britain, London. Available from: https://commons.wikimedia.org/wiki/File:William_Dyce_-_Pegwell_Bay,_Kent_-_a_Recollection_of_October_5th_1858_-_Google_Art_Project.jpg?uselang=fi (accessed November 7, 2016).

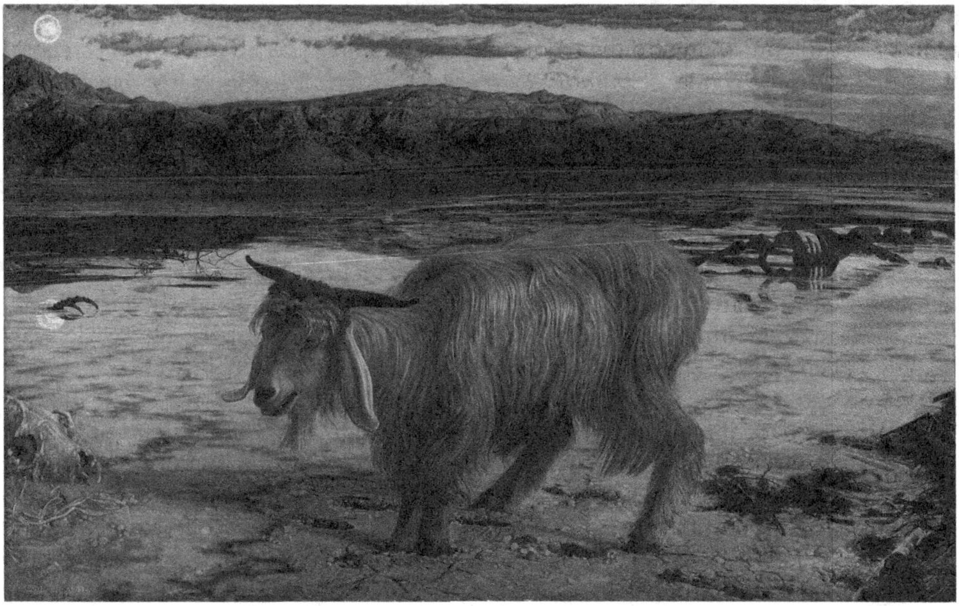

Figure 3.4 William Holman Hunt, *The Scapegoat*. 1854. Oil on canvas, 86 x 135 cm. Lady Lever Art Gallery, Liverpool. Available from: https://commons.wikimedia.org/wiki/File:William_Holman_Hunt_-_The_Scapegoat.jpg?uselang=fi (accessed November 7, 2016).

artwork, for the modernist mind, only if the representation was not reducible to a conventionalized symbolic structure. This opposition easily led to a situation in which the allegoric and the symbolic were read also as some sort of forms of resistance.

To tear down the metaphorical machinery, one needed to overturn it: to repress the abstract meaning and retrieve the material reference, as Norris puts it.[38] But in reality, these positions were not always so clear, and the trends of modern primitivism and objectivist naturalism were often mixed to some degree. As such, the output was not always what one expected. In his own confused way, Pre-Raphaelite William Holman Hunt provides an illustrating example of the situation with *The Scapegoat*.

The Crisis of Symbolism and the Violent Other

While Romanticism thought of nature as a site of human significance, the art of the landscape that was launched by the objectivist paradigm tried to reach a vision of nature that was beyond the boundaries of such processes. But in this sense, the animal was already a problem.

The concept of the animal marked a place of the other also within the objectivist paradigm—a place that, as we have seen, it had already taken within the Romanticist discourse. Only through the new frame did the animal start to redeem the place that it would from time to time acquire within modernist theory: that of an irreducible other that implicated an open resistance against solid significations and conventional symbolism. Paradoxically, then, the need to see organic life in its manifoldness (Darwin's theorem) contributed to the tendency to see animality as an indivisible category of non-humanity (Derrida's "*animot*").

The process was of course quite complex. Animal symbolism had largely been thought already outworn among some of the major currents of thought at the end of the 18th century: the expressionist demands of Romanticism, the cult of individuality, and the belief in art's mission in modern society had changed the nature of the arts thoroughly. Later, there would appear lots of attempts to revive the forms of symbolism—the revisions were often grounded on metaphysical presuppositions, the attempts undoubtedly culminating in French Symbolism. But the crisis had already started, and often the desperate attempts were only a sign that the corpse of the old world was decaying. Henry Fuseli had already spoken of this crisis accurately in 1808:

> In an age of equal refinement and corruption of manners, when systems of education and seduction go hand in hand; when religion itself compounds with fashion; when in the pursuit of present enjoyment, all consideration of futurity vanishes, and the real object of life is lost—in such an age, every exertion confers a benefit on society which tends to impress man with his destiny, to hold the mirror up to life, less indeed to discriminate its characters, than those situations which shew what all are born for, what all ought to act for, and what all must inevitably come to . . . The importance of this object has been so well understood at every period of time, from the earliest and most innocent, to the latest and most depraved, that reason and fancy have exhausted their stores of argument and imagery, to impress it on the mind: animate and inanimate nature, the seasons, the forest and the field, the bee and ant, the larva, chrysalis and moth, have lent their real or supposed analogies with the origin, pursuits, and end, of the human race, so often to emblematic purposes, that instruction is become stale, and attention callous. The serpent with its tail in its mouth, from a type of eternity, is become an infant's bauble; even the nobler idea of Hercules pausing between virtue and vice, or the varied imagery of Death leading his patients to the grave, owe their effect upon us more to technic excellence than allegoric utility.[39]

The crisis was perhaps the most violent in France, where the belief in academic hierarchy of the genres had been the most pressing. Some of the theoretical ideas born in the 18th century we have already traced, and the crisis continued into the next century: the birth of modern art and its need for an absolute vision has long been explained through the dialectics leading to Édouard Manet and the Impressionists, in which the poetics of the Old World were left behind.[40] The dialectics were of course of a truly intricate nature whose every element we can't enumerate. I hope to show in brief, and in light of my examples, that with regard to the status of the animal, one quite important of the currents was not only trying to get rid of the assumedly authoritarian structure of old symbolisms but also of the cold logic of scientific ideas. The animal's place was not, indeed, always on the side of the objectivist natural scientist, not always on the side of the biocentric. "Art and poetry lives through fictions . . . all those things which transport the mind,"[41] Eugène Delacroix wrote. Delacroix's Romantic belief in the powers of art, a belief that scorns all considered scientific and cold, makes an illustrative case of the sort of frame of thought through which the subjectivist worldview could revalue the image of the animal, and of course it has its roots in 18th-century discussions.

In Delacroix, one might argue, Géricault's love of animal imagery—that, in Michael Fried's view, served as an escape route towards a world beyond human reach and the theatrical milieu of old art—turns into a mannerism in which the fierceness of the beasts is valued for its exoticism, for the spectacle of the mythical world of otherness it has to offer the beholder's gaze.[42] And this is indeed what Delacroix said he was after. He stresses the ability of the animal imagery to open a thoroughly fictionalized world in which the scientific world of order, the superficial order of taxonomic language, has been left behind. A passage from Delacroix's journal, dated May 6, 1852, should illustrate the situation. The passage shows how the natural world, even the Romanticist fiction of it, could be represented as a sort of an idealized chaos, thrown into opposition against the vanities of systematic thinking:

> The wise men don't have to live in the countryside, near nature; they like better to influence among the green carpets of the academy, of the *Institut*, of that which all knows like they do; but in the forests and in the mountains, one can observe the natural laws, and one cannot take a step without encountering a subject of admiration . . . The animal, the vegetal, the insect, and the waters are such nourishments for the mind, which wants to study and to register the diverse laws governing the beings. But the academic gentlemen do not find from there a minutest observation that they think worthy of their genius; they want to penetrate further, and to form behind their desks systems that they can take with them to the observatory. Anyhow, they must frequent the salons and to have believers and support; science, which has taken this course, wants any thing but the natural.[43]

"Ah! the birds, the dogs, and the rabbits!" Delacroix burst out, "How these humble professors of good sense, all so silent, all subject to eternal decree, are above our knowledge that is only vain and cold!"[44] The image of the animal—"all so silent"—is here seen as the last means to deconstruct the modern scientific edifice.

By the way, Delacroix's fervent criticism of men of science was in line with his criticism towards the aesthetics of realism.[45] With a sleight of hand, by which the imaginary became equaled with the natural, Delacroix argued that "in the presence of nature herself, our imagination makes the *tableau*."[46] Here, the triumph of human subjectivism is supported by a conception of natural or animal life as the other of human rationality.

The fictional, equaled with the natural, was largely favored in the French Romanticist scene from which the ideology of "art for art's sake" sprang. How was this possible? According to Martin Jay, there had developed within the boundaries of the modern paradigm a dualism that divided the visual register. To be sure, the development towards dualism had started, Jay reminds us, in the Middle Ages, but it could be perfected only after Cartesianism and after the "early modern separation of the visual from the textual . . . which was crucial in the preparation of the scientific worldview."[47] The secular role of the visual led to a principle of observation, and that was what the modern episteme took as its starting point, since the value of objectivity was thus separated from subjective truths on which many of the Romanticist artist and theorists tried still to lean. What was happening at the same time in the artistic field, and which proved to be so efficient for the coming of the new paradigm of art,

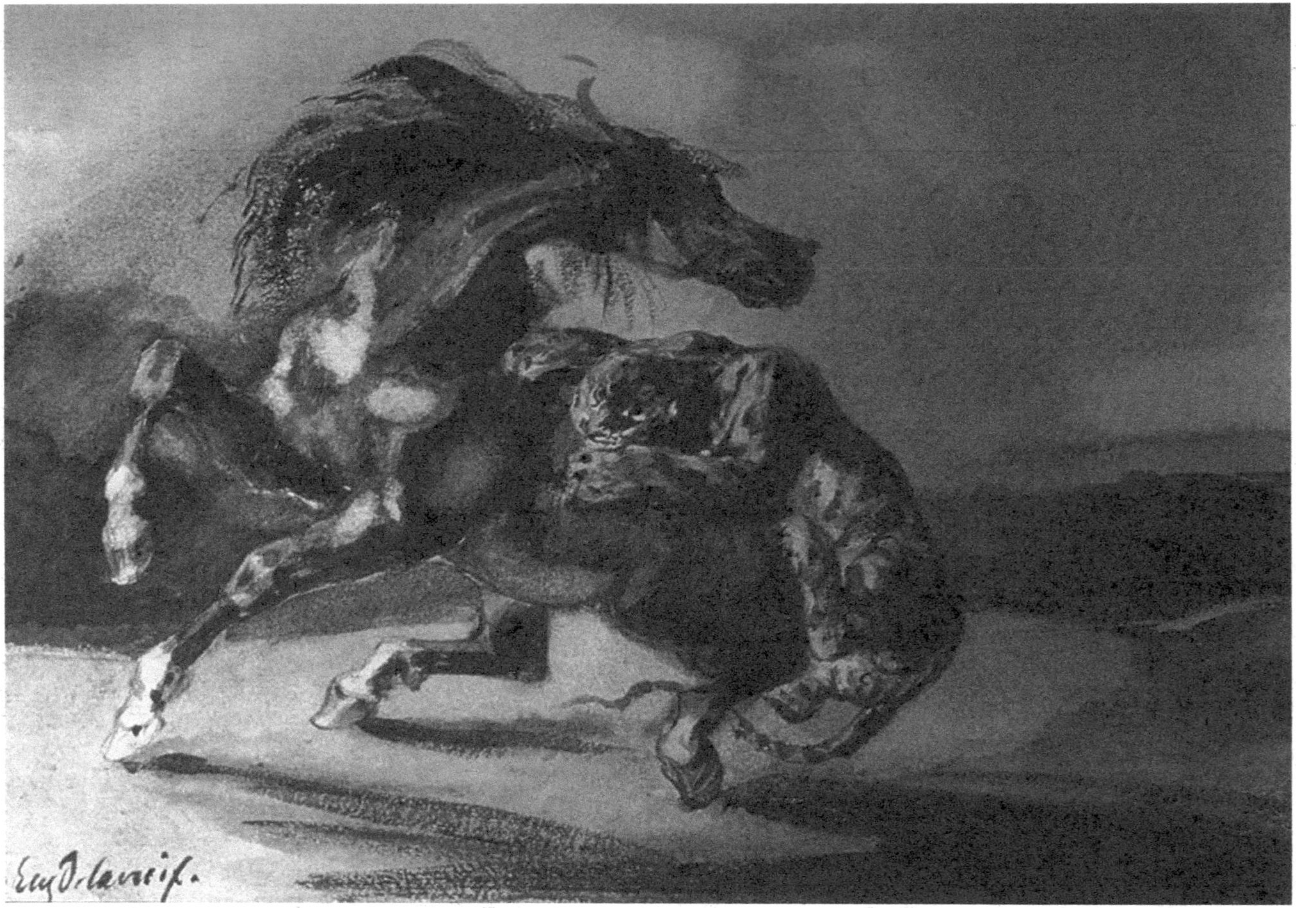

Figure 3.5 Eugène Delacroix, *Tiger Attacking a Horse*. 1825–1828. Watercolor and varnish on paper, 18 x 25 cm. Musée du Louvre, Paris. Available from: ARTstor, www.artstor.org (accessed November 7, 2016).

was the "disentanglement of the figural from its textual task, the denarrativization of the ocular,"[48] as Jay calls it, and to which Fuseli testified with the text just cited. The book of nature, or, the classic symbolic order, did not survive the scientific and social revolution.

The dream of an objective vision, in its ideality, itself produced the dualism of the modern vision. There was always the ambiguity inside human observation that was a result of the desire inherent in seeing. And under the sign of this ambiguity, the imagination was seen opening towards the unknown, and the irreducible otherness of the animal subject started to acquire new meanings. "For if vision could be construed as either the allegedly pure sight of perfect and immobile forms with 'the eye of the mind' or as the impure but immediately experienced sight of the actual two eyes," Jay writes, "when one of these alternatives was under attack, the other could be raised in its place."[49] Ever since, dualism has divided artists into two groups: either for or against the "retinal shudder" which horrified Duchamp[50] and that the Impressionists embraced. For the same reasons, Delacroix despised realist and naturalist efforts; they were reducing his art to a work of science: pure, objectified seeing.[51] But regardless of the side a given artist took, Jay reminds us, "something called vision could still be accounted the noblest of the senses."[52] This might also explain the autonomous value that the visual arts were given under modernity, at least in appearance.

This leaves an even greater ambiguity at the heart of the ways the image of the animal could be evaluated under the dualist structure. And in this regard, it is not surprising that it was the side that was for "art for art's sake" that could still embrace images of animals. Animal images not only carried within themselves the ancient load of symbolism, which the "objectivist side" would then have to get rid of one way or the other, but they also opened a view towards a world that truly was outside human measures and presumably imaginable only through fiction. There was now always this scent of fiction, this surplus, in the images of animals.

Delacroix's ideas on the admirableness of animal subjects[53] seem indeed to have had some resonance even within the stiff practices of the French salons. While in the 1830s Antoine-Louis Barye, as a foremost *animalier*, was rejected several times due to the subjects of his sculptures being mainly "only" animals,[54] in 1853 Henri Delaborde was already noticing that there were "today almost as much artists who are dedicated in studying animals than those studying human beings."[55] Although there were of course as many conservative voices as the others, it may serve as a sign of the times that the Brothers Goncourt could even come to claim that animals were replacing man in the evolution of modern art.[56] Especially, they thought, this was happening in sculpture:

> At this historical moment, the art of sculpture is going through a change that has already shown itself in the art of painting. Historical school is dying for art that makes palpable in sculpture, and for art that makes visible in painting. Landscape replaces historical school in painting; animals replace it in sculpture. Nature succeeds man. Such is the evolution of modern art.[57]

Thus, the question was not so much about conceiving animals themselves in a more empathic light; at stake was a larger theoretical development through which the hierarchy of the visual arts was contested. In other words, these developments had less to

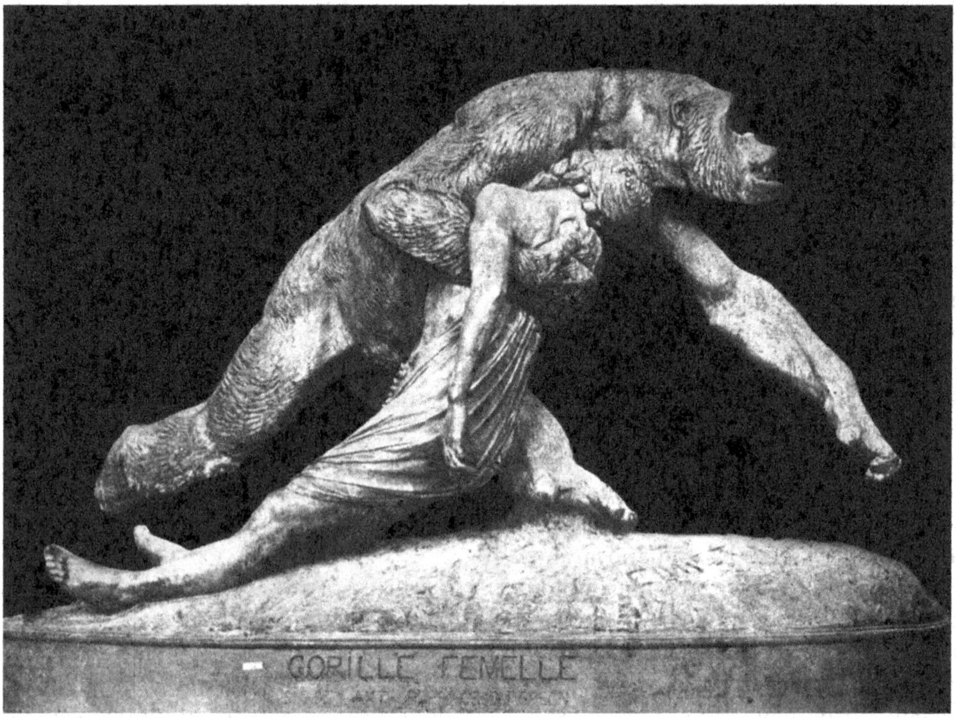

Figure 3.6 Emmanuel Frémiet, *Gorilla Carrying Off a Negress*. 1859. Plaster, Greater than life-size (destroyed in 1861). Available from: https://commons.wikimedia.org/wiki/File:Gorille_enlevant_une_négresse.jpg?uselang=fi (accessed November 7, 2016).

do with animality than with the nature of the allegory and of the symbolic quality of the images. Rather, the imaginary was now often in the service of the violence inherent in the downfall of the system of signification.

For example, when in 1859 Emmanuel Frémiet sent to the *Salon* his work *Gorilla Abducting a Negress*, it produced outrage—after heated discussions, a part of the audience was admitted to see it but only behind curtains.[58] The theme of the work, foreshadowing *King Kong*, was actually copied from an illustration from Linnaeus's *Universal System of Natural History* from 1795,[59] and it showed an orangutan carrying a woman away. However, when a new version of Frémiet's orangutan was exhibited in the 1887 *Salon des artistes françaises* it achieved a *Salon Medal of Honour*, and it was also later shown in the 1889 *Exposition Universelle* as part of the French State Collection.[60] Something had truly changed in the attitudes towards the depicted scene; but perhaps the scene itself was not as important as the readings that determined its content. In 1859, Charles Baudelaire found, quite tellingly, the value of the work, even without seeing the whole sculpture, and openly confessing the fact, in the obscenity of its topic:

> The orangutan carrying a woman into the woods (a refused work, which naturally I haven't seen), is very much the product of a sharp mind. Why not a crocodile, a tiger, or another beast likely to eat a woman? Why not? Remember, he has no interest in eating her, but in raping her. For the lone ape, the gigantic ape, at once more and less than a man, has sometimes shown a human appetite for woman.[61]

In Baudelaire's reading, the emphasis is, on the whole, on one essential characteristic: the gorilla/orangutan is not an equivalent to the bull that raped Europa, it is not an allegorical figure living in a mythical space but an animal that was supposed to be taken as an animal and nothing else. In Hegel's terms, it has lost its qualities of an "abstract attribute."[62] The shock value of Frémiet's work is in it: Frémiet had indeed been on the track of finding an expression to a materialist vision that would give his animal art a value in itself in the hierarchized French art world, and we also know of Frémiet's influences that he was attracted to the theories of natural history and evolution, and tried to adapt their principles to his art.[63] To put it more straightforwardly, its theme was not an allegory—a work depicting the manners of men and gods through animal imagery—but constructed on a slight difference between the species depicted, the human and the ape, an essential difference that swallowed up the whole work. Whatever the aims, Baudelaire's reading of the work emphasized, still as anthropocentric as ever, that the animal showed the negative side of humanity, but humanity now lost in its own violent animality: "a human appetite for woman," as Baudelaire put it.[64] Of course, there were lots of possible ways to interpret the image, but even the possibility of such an interpretation Baudelaire was after showed the way—and despite his evident mistakes, not *completely* in the wrong direction.[65] The principal question concerned the ways such animal statues could be interpreted given the crisis in animal symbolism.

The point that I try to emphasize with the exemplary case of Frémiet is that in 1859 there was still almost a total lack of means to deal with such visual attempts as those Frémiet was after—an animal statue on the brink of a narrative structure but stripped of its conventional symbolic contents—while in 1887 the means to interpret the work had developed much further (the latter version is evidently also the more obscene). The earlier one could be interpreted mostly through the human behavior that marked the sphere of the other in the civilized Western world—a condition, as we shall soon see, that attracted many of the early-20th-century avant-gardists towards the philosophical concept of the animal.

All this marks the concept of human animality for which the 19th century, after Michel Foucault, has become infamous. Madness, aggressive impulses, apathy, hysteria—the attributes of unsocial behavior had over the last centuries been more and more attached to this certain kind of human animality.[66] The feminine animal—or the animalized female—of the Symbolists is a well-known case in point.[67] For the moment, the image of the animal was not becoming only anti-allegorical but sociological as well.

It was this path that had already been foreseen by Fuseli. If the image of the animal stayed at the core of a world where fiction ruled, but where it had lost its symbolic tasks, it could only have significance and maintain itself within an anthropocentric culture by turning into a figure of irreducible otherness: "infant's bauble" or a subject of our worst nightmares.[68] Nightmares, certainly, of ourselves.

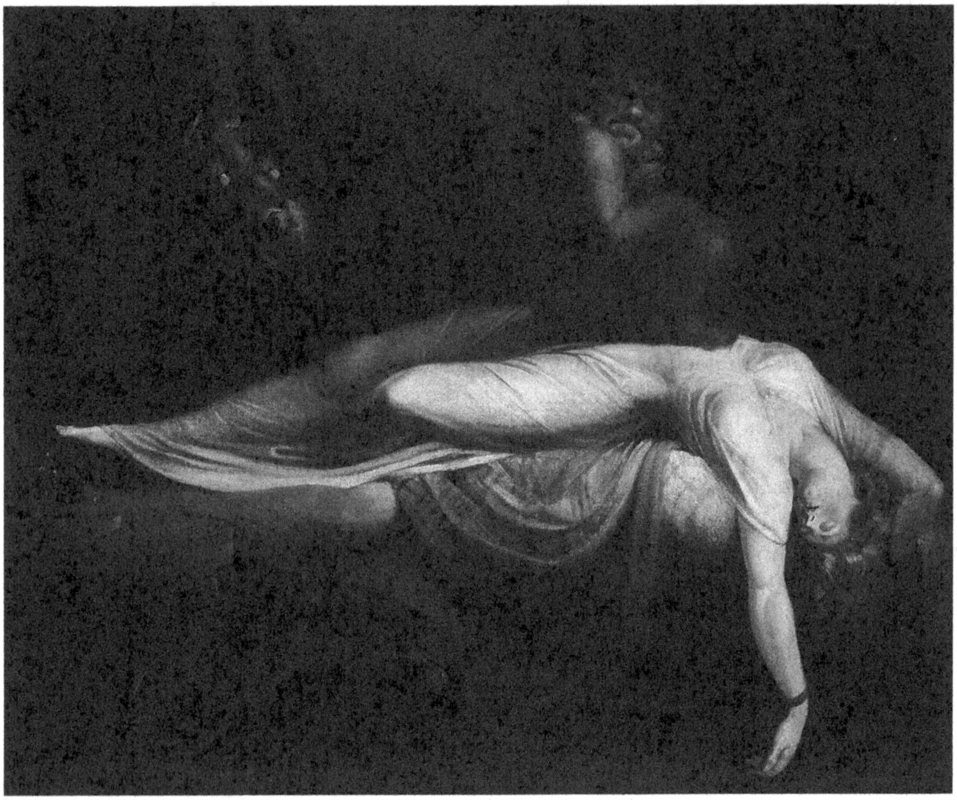

Figure 3.7 Henry Fuseli, *The Nightmare*. 1781. Oil on canvas, 101.6 x 127 cm. Detroit Institute of Arts, MI. Available from: https://commons.wikimedia.org/wiki/File:John_Henry_Fuseli_-_The_Nightmare.JPG?uselang=fi (accessed November 7, 2016).

Evolution of the Species and the Creative Sentiment

As suggested, the opposition between the subjective and the objective paradigm was seldom complete within individual artworks. And even the possibility of thinking of this opposition in that way imposed a problem: artistic creativity, anyhow, could not be represented through scientific reductionism. Neither was the animal capable of resisting endlessly; its ability to represent the unrepresentable, the world outside human measures, was soon understood as a symbolic value in itself.

Into this field, Darwinism and the different forms of naturalism were implanted. The task was not easy but seemed somehow necessary. Darwin himself had proposed that man's cultural expressions could be studied within the boundaries of natural evolution,[69] but the quest itself faced almost insurmountable problems, since the practice of art was an exclusively human purview, even if it was granted that the principle of creativity could be considered as belonging to animal life. How could, then, the origin of art and art's inner nature be explained in terms of the argument maintaining that the cultural could be reduced to the natural in its entirety? And if one did not scorn art—as the theories of it still quite rarely did—it was often found that one needed to back up the naturalist argument by proposing some sort of fixity with regard to the human-animal boundary, despite the aims of one's discourse. Sometimes, the naturalists did not need to discuss the subject of the animal at all: if the perspective posited animals and humans on the same level, the concept of the animal needed no attention in itself. One such current could be found in the field of French Naturalism: since its interpretation of the evolutionist frame was mainly adopted from the works of Jean-Baptiste Lamarck and Étienne Geoffroy Saint-Hilaire, whose theories differed from Darwin's mostly because of the emphasis that the two laid on the interplay between the will and the environment, while the Darwinist side understood the whole development more easily through its ideas on natural selection.[70]

Towards the end of the 19th century, the evolutionist paradigm was in many ways somehow consolidated with the Romanticist love of artistic creation, though the solutions varied to a large extent. Within these efforts, the animal also found its new place within the philosophies of art. Forced as some of the theories were, their positions and ideas were highly understandable within the context in which the theory of art also had to react to the developments in the sciences of its time. One needed to form a synthesis of the radical tendencies to separate between creativity and nature, and often the propositions, rarely less radical in their views, contributed to the worsening crisis of symbolism.

One of the most influential solutions of this kind was provided by Ernst Haeckel's *Darwinismus*. Mostly known as a biologist and as a naturalist philosopher, Haeckel also contributed to the art theory of his time. He was interested in building what he called *Natürliche Schöpfungsgeschichte*[71]—ultimately, an expanded version of natural history that would settle the unanswered questions of the entire evolution of organic life by synthesizing it with the philosophies of the German Idealistic tradition.

Haeckel's central idea focuses on a twofold view of creation history, a view that combines fixist and evolutionary ideas on the human-animal divide. Firstly, although Haeckel situates man in the territory of natural evolution, he does this by maintaining that mankind alone populates the highest, 22nd step of the evolutionary ladder—the 21st step being the "missing link" between apes and humans that had more or less been theoretically sought since the 18th century. Haeckel's idea of the missing link indeed focuses quite traditionally on the essential difference between man and animal, proven by man's ability to speak: the hypothetical missing link is considered *Pithecanthropus alalus*, "man without speech."[72] However, Haeckel's theoretical construction does not limit itself to the question of speech, it maintains a real continuity between man and the history of natural creation. This side of the theory is represented by his well-known proposition—later embraced, for example, by Freudianism—that "ontogeny recapitulates phylogeny," a theory known as the recapitulation theory. Despite the fact that the theory is nowadays largely discredited, it has been quite influential and it also stamped the art theory of its time. The view consists of the idea that in developing from embryo to adult, human beings, as well as other animals, travel through all of the successive evolutionary stages that have molded the current form of their species. In this sense, Haeckel also underscored that animals and men were made of the same elementary particles, even in a spiritual and cultural sense. According to his monist principles, he was even ready to think of souls as only "a sum of a number of special cellular activities, among which feeling and will, sensation and involuntary movement are the most important and the most generally distributed."[73]

Anyhow, if this vision led Haeckel to maintain that both cells and protozoa took part in soulful activity,[74] there was an assumed difference between the lower and the higher

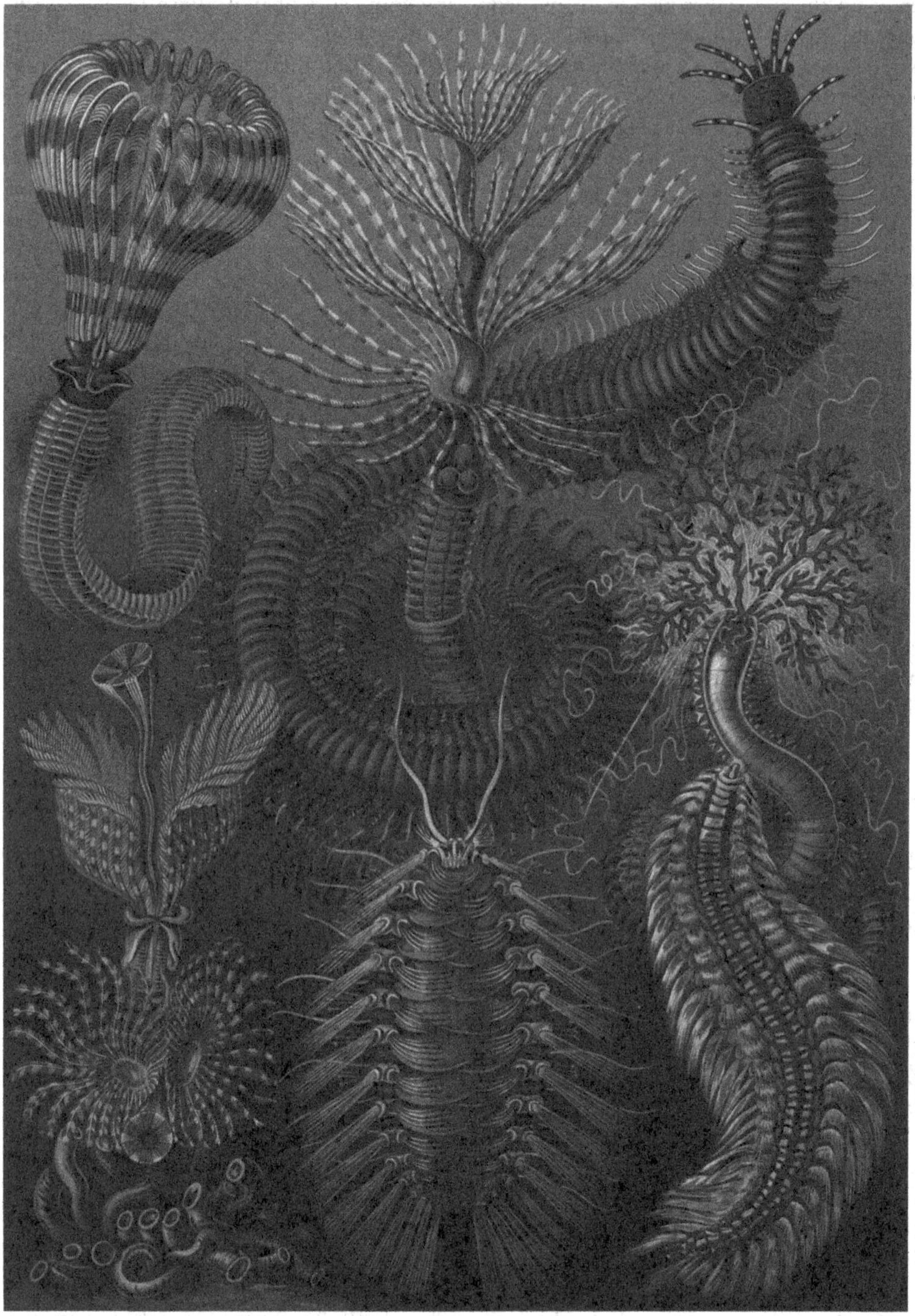

Figure 3.8 Ernst Haeckel, *Kunstformen der Natur, plate 96 "Chaetopoda."* 1904. Scientific illustration. Available from: https://commons.wikimedia.org/wiki/File:Haeckel_Chaetopoda-edit.jpg?uselang=fi (accessed November 7, 2016).

stages of creation. He insisted that while cells possessed a "cell psyche" (*Zellenseele*), men and higher animals possessed "psyche cells" (*Seelenzellen*), the latter making functions like thought, representation, and human speech possible.[75] Ultimately, the monist principle was given away in favor of a conservative philosophy that asserted that the higher animals were such only because of their representational abilities. The ability to represent—and this was what lured the artistic community—acted also as a proof of a possibility of a recollection of the more-or-less psychological vibrations experienced in the former stages of individual life and creation history.[76] This meant that Haeckel's *Darwinismus*, as it was called, influenced as it was by the German *Naturphilosophie* of its time, could also be used to motivate a certain form of modern primitivism. This form relied upon the old idea whose validity was supported by the traditions of Romanticism and Hegelianism: the idea of art's capacity to open up man's consciousness to the totality of one's own history, of man's consciousness of self. The difference was that the way now passed through primordial animal sentiments, even those acquired when we once lived in the sea.[77]

Darwin had indeed suggested that our life too had its origins in the sea; the hypothesis soon developed, within Haeckel's system, into an idea that the animals closest to

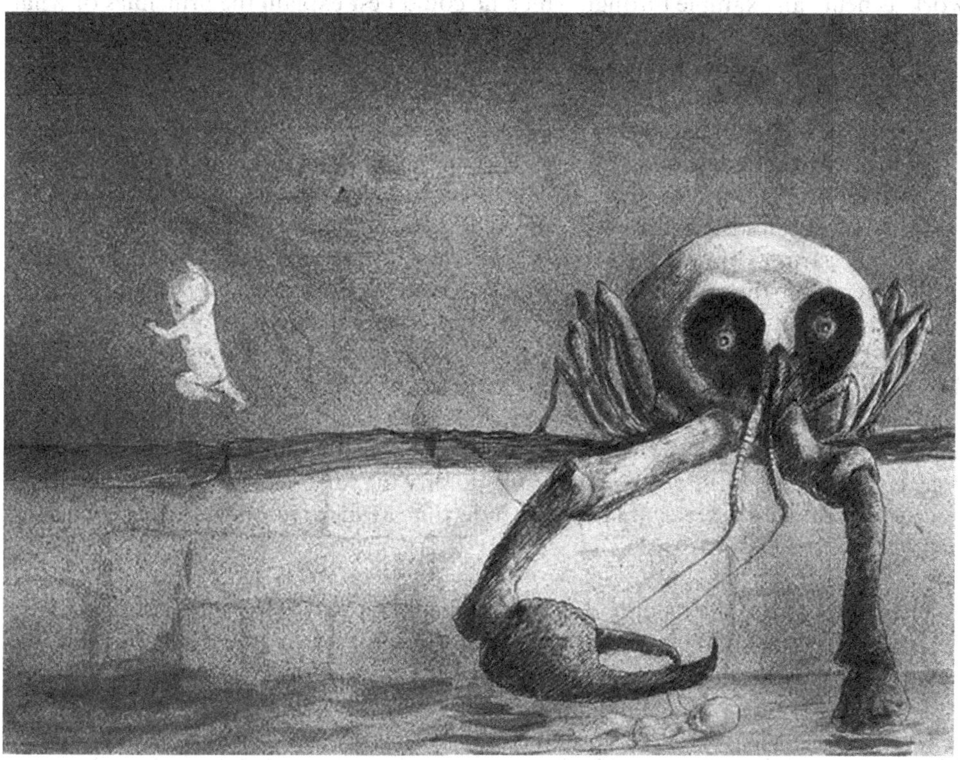

Figure 3.9 Alfred Kubin, *The Moment of Birth*. 1902. Facsimile print on japan paper, 22.9 x 31.8 cm. Available from: www.wikiart.org/en/alfred-kubin/the-moment-of-birth-1902 (accessed December 2, 2016).

the primordial stage, the *monera*, were the ones from which the unified experience of the vibrations of the protoplasm—understood as "a primordial slime" [*Urschleim*], which was seen as the basis of life—could be examined more precisely. From this, it in turn followed that the sea started to serve as a model for primordial life and experience, an experience of the unification of the sensual impressions, an idea evidently effectuated by the Romanticist views on the experience of poetry.[78] If the sea indeed started to replace Eden for Darwinists and their followers, as Marsha Morton suggests,[79] the development also left the sea in a role in which it once again could freely be represented as the antithesis of man's reasoning, and posit itself as the object of the horrors we have of the origin.[80] The sea had become, according to David Kelley's humorous phrase, an ideal place which "tapped the primeval memory of your body linking you to the primitive wild love orgies of bygone eons."[81] The artists could now let their imaginations flow between the devil and the deep blue sea. Alfred Kubin's art could exemplify the tide.

The resonances of these developments were mostly felt in the German-speaking countries, most notably in the tradition leading from German Symbolism to Expressionism. One of the prominent propagators of Haeckelian ideas was the Polish-born writer Stanislaw Przybyszewski, influential among the Symbolists and closely associated with August Strindberg and Edvard Munch. Przybyszewski thought of Munch's work as being an example through which he could best explain the principles of what he called "psychic naturalism." Quite tellingly of the twofold relationship that some of the Symbolist theories had to the concept of the animal, Przybyszewski wrote that:

> [T]he most highly developed form of life is man himself. It is like a great wave which grows everlastingly, a germ which perpetuates itself to infinity in constantly changing metamorphoses, and thus in every individual human being this individuality constitutes the collective focus of all those features which have characterized all the various branches of its own previous evolutionary development: it is a pangenesis in the sense understood by Darwin: every original generating cell carries the whole of human kind and all its characteristics within itself.[82]

The origin of this sort of artistic creativity was on the irrational side, that is to say, on the animal side. At least, this was the case as long as animality could be represented at the same time as the core of our being, as well as a condition of otherness with regard to our reasonable ways of life. Such duality made possible some nice rhetorical figures; Przybyszewski was able to conclude that Munch, in his assumed greatness, had become the first artist

> to represent the most subtle and inconspicuous of psychological processes just as they appear spontaneously . . . quite independently of any mental activity on our part . . . when all conscious activity of the imagination has ceased to function: distillations of the animal and irrational soul.[83]

The passivity of the animal soul had curiously once again made its way into thoughts about the nature of the arts.

The alternative way to support the Haeckelian idea was to emphasize the sensual qualities of the experience of art instead of the primacy of the artistic intuition in the formation of man's consciousness of self. The option was mostly embraced by the French Symbolists, who had already taken Arthur Rimbaud's demand—presented in the famous *Lettres du voyant*—of the "deregularization of all the senses" ["dérèglement de tous les sens"][84] as a value in itself. The Haeckelian model responded to ideas about the synesthetic qualities of the true arts; the Haeckelian dream about the primitive animal, able to feel and sense without the differentiation of her sensual faculties appealed to the minds of the theorists who tried to synthesize between the demands of scientific thinking and art theory.[85] One proof of the excitement possessing the minds of the Symbolists, was an observation made by Raphaël Dubois: the physiologist had been astonished when finding that the piddock (a type of mollusk) had only one organ through which the different sensual impressions were received. Dubois himself had proceeded in describing the life of the mollusk as "dermoptique"—as determined by "touch-vision"—and contrasting it with the intellectualized vision of humanity.

The prevailing thoughts on art and *synaesthesia*, in a typically French manner, connected the nature of the visual arts once again with animality, although still mostly affirming the exclusively human nature of artistic activity.[86] The origin of these sensual experiences, artistic in their nature, was deduced from theories about primordial vibrations. To give an example, another prominent physiologist, Charles Henry, who had a legion of followers among the Neo-Impressionist circles, had an idea of the artist who could siphon the vibrations through his human faculties. Henry claimed that "the artist is but an eye, an ear, a nervous system normally organized and developed" that senses a rhythm out of an "infinity of more or less invisible rhythms," realizes it virtually as an "Idea, and produces it externally."[87] It did not take long for one follower of Henry's, Jenny d'Héricourt, to understand that he had found there "a veritable mechanics of the protoplasm as revealed."[88]

The views were not without problems. The different arts usually appealed primarily to only one sense, and one had to choose sides. In a post-Impressionist context it was often the eye—and the qualities related to vision—that were somehow made central. An influential scientific figure close to the avant-garde, Félix Le Dantec, stood at the watershed in this regard. Lamarckist, positivist, anarchist, and an atheist as he was, he supported the artistic theory that claimed that art was connected with our animalized "ancestral influences,"[89] but he also criticized the vulgarity of the synaesthetic theory then prevailing. Dantec's view consisted of the contrasting idea, maintaining that in higher animals one of the senses always superseded the others, the one integrating the sensations received from the others. Le Dantec—in 1901, near the anarchist avant-garde that kept faithful to the positivist principles so far as they suited its own political premises—supported the primacy of vision with his theories: "There is a capacity, and a drive, to reunite the senses not in synaesthesia per se, but in a unification of the senses under the direction of vision," he wrote.[90] The protoplasmic condition could be achieved again only through human means, since the human eye would now assure Le Dantec's readers of man's place at the top of the ladder of creation history.[91]

The Darwinist tendencies were indeed quite easily connected with the political avant-garde, and with such connections both were also criticized for the failures of the other.[92] As always, the primitivist tendencies generated a reaction—for some, the elevated aim of acquiring higher consciousness of self did not qualify as a reason. Hippolyte Taine, among others, while acting as a professor of aesthetics at the *École des Beaux-Arts* in Paris, condemned the Darwinian development in the field, at least politically, when he used Darwin's *The Descent of Man* as an illustration of the degeneration of the Paris Commune.[93] And Max Nordau, to give the most notorious example, thought that the theories in circulation were largely evidence of a degenerate culture. Nordau referred especially to the French discourses on the topic by claiming that in his view:

> [T]o raise the combination, transposition and confusion of the perceptions of sound and sight to the rank of a principle of art, to see futurity in this principle, is to designate as progress the return from the consciousness of man to that of the oyster.[94]

A condition, of course, which the theories in question were after indeed, at least apparently. The most guilty party in this sense, according to Nordau, were the Symbolists, because of their appreciation of the irrational and ritualistic, and who, in Nordau's view, had "activity of the organic nerves that preponderated over the cerebral cortex, their emotions overruling their ideas."[95] "When persons of this kind have poetic and artistic instincts," Nordau concluded, "they naturally want to give expression to their own mental state."[96]

In one way or the other, the animal condition was now highlighted; it now offered itself not only as a goal to reach through artistic means (as in some of the Haeckelian theories) but also served as an artistic identity. An avant-gardist artist could from now on explain his identity in animalized terms—"bête comme un peintre,"[97] as the saying went—or at least appropriate Darwinist worldviews, while constructing her own identity as a genius. Thus, some of the Symbolists were adopting the perspective that Nordau's vision opened in front of them. Edvard Munch, for example, thought of himself as the degenerated artist *par excellence*, whose manners could testify to his status as a loser in the combats of life and of sexual selection but as a winner in the artistic arena.[98]

Nietzsche

Before ending my discussion of the theoretical relation of the concept of the animal and the arts under the influence of Darwinism in the 19th century, I would like to introduce a figure whose contribution to the theme at hand is measureless, Friedrich Nietzsche. Nietzsche's contribution is certainly an ambiguous one, but it opens a perspective that became effective among the avant-garde in the first decades of the 20th century. One might even say that it revolutionized the whole field, but its merit was also to merge into itself many of the currents we have been following, however conflicting they may have seemed. Nietzsche managed to do this when he put the concept of the animal at the center of his thinking about the arts.[99]

To begin with, I argue that Nietzsche's criticism appropriates three important themes, all already known to us:

(1) The thesis of human exceptionality, which ensures that the animal can be equipped with the status of the irreducible other. Within Nietzsche's thought, the idea is introduced only through a rigorous critique of human self-consciousness.

(2) A critique of conventional symbolism incorporating into itself a critique against the whole of Western culture—in Nietzsche's terminology, civilization—namely, against the forms of systematic thinking that it offers.
(3) A possibility of a total experience, a Romanticist idea of the arts, that is transposed into a post-Darwinian frame of reference—an idea of an artistic experience capable of mystically crossing the line between a human subject and its object.

As suggested, Nietzsche's "ego-mania"[100]—in Nordau's terms—grounds itself once again in a critique of the falsity of man's self-consciousness. This, to be sure, it does with an ambiguous gesture. His famous formulations of human beings as the "clever animals that invented knowledge,"[101] or as the "clever animals that had to die,"[102] are even more equivocal than they seem to be at first sight. They both implicate the old argument that humanity, despite its belonging to a genealogy of animals, is different from the other animals due to its reason, self-consciousness, and knowledge of death—despite the sardonic tone. A larger consideration of the framework in which they appear should give us a clue to his elaborations on the theme. The first of the citations, taken from Nietzsche's "On Truth and Lying in a Non-Moral Sense" is from a paragraph that links Nietzsche to the objectivist paradigm, that is, to the demand to see outside the limitations of human perspective. This is how Nietzsche conceives of the exceptionality of our species:

> In some remote corner of the universe, flickering in the light of the countless solar systems into which it had been poured, there was once a planet on which clever animals invented cognition. It was the most arrogant and most mendacious minute in the 'history of the world'; but a minute was all it was. After nature had drawn just a few more breaths the planet froze and the clever animals had to die. Someone could invent a fable like this and yet they would still not have given a satisfactory illustration of just how pitiful, how insubstantial and transitory, how purposeless and arbitrary the human intellect looks within nature; there were eternities during which it did not exist; and when it has disappeared again, nothing will have happened. For this intellect has no further mission that might extend beyond the bounds of human life. Rather, the intellect is human, and only its own possessor and progenitor regards it with such pathos, as if it housed the axis around which the entire world revolved.[103]

The faculty of self-consciousness is not affirmed as the quality that makes man different from the other animals, rather, the opposite. Nietzsche is quite clear, reminding us of Diderot's arguments, that the egocentric way of being in the world belongs just as well to nonhuman animals: "But if we could communicate with a midge we would hear that it too floats through the air with the very same pathos, feeling that it too contains within itself the flying centre of this world."[104] It is not hard to see why some commentators on Nietzsche have thought that he goes so far as to annihilate the thesis on human exceptionality—sometimes the argument is supported with reference to Nietzsche's Darwinism.[105] Nietzsche, however, maintains throughout his writings an idea of human beings as different from the rest of the animal kingdom just because man keeps fantasizing of himself as being the crowning achievement of creation. This principle Nietzsche states with an ultimate clarity in a paragraph in *The Gay Science*:

> *The four errors.*—Man has been educated by his errors: first, he saw himself only incompletely; secondly, he endowed himself with fictitious attributes; thirdly, he placed himself in a false rank order in relation to animals and nature; fourthly, he invented ever new tables of goods and for a time took them to be eternal and unconditioned—so that now this, now that human drive and condition occupied first place and was ennobled as a result of this valuation. *But if one discounts the effects of these four errors, one has also done away with humaneness, humanity, and 'human dignity'.*[106]

The same proposition is elicited in another context, but now from "an animal perspective":

> *'Humanity'.*—We do not regard the animals as moral beings. But do you suppose the animals regard us as moral beings?—An animal which could speak said: 'Humanity is a prejudice of which we animals at least are free'.[107]

From this perspective the maintenance of the difference, or the transgression of it, mainly matters in Nietzsche's thought. In other words, the falsity of man's self-consciousness, following Nietzsche's ideas, opens for him the possibility of bringing into light a view of an idealized cultural sphere connected with animality true to itself.

As odd as the idea at first may seem, it gets its force from an antagonism that Nietzsche sketches between culture and civilization. This opposition was not new, though it had not usually been posited with such severity. Civilization, an 18th-century French term denoting high culture, had been devalued in German thought for the whole century: in Romanticist circles, the term had been understood as the human civilization and its practical and technical ways of dealing with nature, when the cultural (*Kultur*) was thought as being the sphere of the spirit; the same situation was renewed in a Marxist context in which the civilization became known as the infrastructure ("roads, ports, and quays"), while culture equaled with the superstructure ("science, art, order, and virtue").[108] Nietzsche, however, made out of these dualities an antagonism that he could make use of in his critical texts. At the same time, his "philosophy of life" modified the common meanings of the two terms. This is how Nietzsche understood their relation:

> The highpoints of culture and civilization do not coincide: one should not be deceived by the abysmal antagonism between culture and civilization. The great moments of culture were always, morally speaking, times of corruption; and conversely, the periods when the taming of the human animal ("civilization") was desired and enforced were times of intolerance against the boldest and most spiritual natures. Civilization has aims different from those of culture—perhaps they are even opposite.[109]

To put it otherwise, the taming that is called civilization is the principle of human alienation, which is evidenced by the "fictitious attributes"[110] with which man endows himself. From this it follows that man—who according to Nietzsche, "occupies the same level of perfection" as animals—is for him "in no way the crown of creation," but is still unique because Nietzsche thinks that men are "the sickliest animals, who have strayed the most dangerously far from their instincts."[111] Thus, he confirms, men are also "the most interesting" of all of the animal species.[112] Man may belong to the same plane of life as all the other animals, but is also different from them because of his essentially civilized nature.

As Keith Ansell-Pearson has noted, the main difference between Nietzschean philosophy of life and Darwinism is perhaps buried behind Nietzsche's view that the difference between man and animal is found on the cultural level: Nietzsche's "philosophy of life" tries to maintain that man's futile uniqueness is a matter of the relative differences between human and animal cultures. Ansell-Pearson aptly remarks that while Darwin tried to think of human evolution in terms of natural selection, Nietzsche considered it a process of cultural selection.[113] Further on, Nietzsche's perspective, as Vanessa Lemm has emphatically written, makes it possible to observe "the dual questions of what it means for an animal to have culture and how animality engenders culture."[114] The kind of unique form of alienation—that is, civilization, in Nietzsche's terms—also famously led him to define humanity, in *Thus Spoke Zarathustra*, as something that "must be overcome."[115] In the context of our research, the idea can as well be equated to Nietzsche's demand for the animalization of human culture: "the human being is" only, as Nietzsche's cultural politics propose, "the *not yet confirmed animal*"[116]—alienated from himself as man is. "Zarathustra's animals," the eagle and the serpent, lead him in his spiritual path.

The animal perspective kept Nietzsche somewhat busy within the boundaries of his project of "the revaluation of all values," in relation to the questions initiated by aesthetic discourse. One of these values to be reassessed was evidently the human conception of beauty: "beauty in itself" Nietzsche judged to be "an empty phrase, not even a concept."[117] Although Nietzsche can without a doubt be thought to represent the German Romanticist tradition because of the primacy he attaches to the revolutionary dynamism of the arts, it is also in this sense that he tries to surpass the limitations of the tradition. In *Twilight of the Idols*, Nietzsche responds for his predecessors by demanding a total forgetting of the principle of human beauty. The tone of the demand may sound familiar in the light of our exposition of Darwinist thought, but its consequences will be quite different:

> *Beautiful and Ugly*. Nothing is more highly conditioned—let us say: more *limited*—than our feeling for beauty. Anyone trying to think about this feeling in abstraction from the pleasure human beings derive from humanity will immediately lose any sense of orientation. "Beauty in itself" is an empty phrase, not even a concept. In beauty, human beings posit themselves as the measure of perfection; in select cases they worship themselves in it. In this way, species cannot help but say yes to itself and only itself. Its *lowest* instincts, those of self-preservation and self-propagation, shine through in sublimities like these. People think that the world itself is overflowing with beauty,—they *forget* that they are its cause. They themselves have given the world its beauty—but oh! only a very human, all-too-human beauty . . . Fundamentally, humanity is reflected in all things, people find beauty in everything that throws their image back at them: the judgment 'beautiful' is the *vanity of their species* . . . Of course a sceptic might hear a suspicious little whisper in his ear: does the world really become beautiful just because it is seen that way by human beings, of all creatures? People have *humanized* it: that is all. But nothing, absolutely nothing,

> guarantees that a human being is the standard of beauty. Who knows what a human being looks like in the eyes of a higher arbiter of taste? Daring, perhaps? Perhaps even funny? Perhaps a bit arbitrary?[118]

In relation to the matters of art and aesthetics, the quest becomes then to forget the measures that gave to the humanized principle of beauty its value: the measures through which man measures, or had measured, himself. The revaluation of the principle of beauty would then lead to a culture that, by means of animal otherness, had forgotten the measures of human civilization. Thus, Nietzsche sees that animality could be surveyed through the forgetfulness that it represents. His view may lead to an ideal projection of the otherness of the animal, a projection that only repeats the old arguments about animal nature, but as such animality acquires new value in Nietzsche's cultural politics:

> This is a hard sight for man to see; for, though he thinks himself better than the animals because he is human, he cannot help envying their happiness—what they have, a life neither bored nor painful, is precisely what he wants, yet he cannot have it because he refuses to be like an animal. The human being may well ask an animal: "Why do you not speak to me of your happiness but only stand and gaze at me?" The animal would like to answer, and say "The reason is I always forget what I was going to say"—but then he forgot this answer too, and stayed silent: so that the human being was left wondering.[119]

In this sense, the concept of the animal (the *animot*) becomes a sort of a symptom of the discourse—a symptom that marks a possibility of an organic life outside humanity and its restrictions. In other words, animality carries within it a possibility of unraveling the vicissitudes of human civilization. Partly, this is a consequence of Nietzsche's subjectivism: the phantasmic return to one's own nature—an equivalent of the eternal return—does not subserve any Rousseauesque dream of the primitive, but gives Nietzsche the chance to stand against the constraints of civilization, "to educate oneself . . . against one's self, to a new habit and nature, out of an old first nature and habit."[120] The "second nature" is, then, brought up by the revolutionary culture that is posited by human animal's subjective singularity.[121] According to Nietzsche, animal "conceals [*verbigt*] nothing and at every instant appears [*erscheint*] wholly as what it is; it can therefore never be anything but honest,"[122] while men have always falsely "presented depersonalization as the cause of the greatest knowledge and ability."[123]

Thus, the animal's healthy way of egoistic thinking, to put this in Nietzsche's own terms, lies in its silent resistance. The usual understanding of the human-animal divide still maintained in Nietzsche's era—that conceptual thinking is an exclusively human faculty—is here introduced as a frailty. At the same time, Nietzsche, however, adopts the view supported by Delacroix that our fantasies about animal nature can be made to attack systematic reasoning, the cause for man's alienation.

All this leaves open the question of method. How, one might ask, can one posit through human culture one's own animality and leave one's own nature for the sake of the "second nature," which is "more natural than the first"? Nietzsche's proposition is that the value inherent in animal-like behavior is in the suspension of logic that "animal forgetfulness" brings about. The suspension, similar as in a dream, makes up the connection between animals and art that Nietzsche enshrines. "Art reminds us of conditions of animal vigor," he states:

> it is sometimes an excess and effusion of burgeoning corporality in the world of images and desires; at other times, an arousal of animalistic functions through images and illusions of intensified life, a heightening of the sensation of life and its stimulant.[124]

What is left behind in these "images and desires" is the first delusionary form of forgetfulness, not the authentically animal one, but the civilized way of forgetting that has led to man's disdainful nature:

> Only by forgetting this primitive world of metaphors, only by virtue of the fact that a mass of images, which originally flowed in hot, liquid stream from the primal power of the human imagination, has become hard and rigid, only because the invincible faith that *this* sun, *this* window, *this* table is truth in itself [*Wahrheit an sich*]—in short only because the human being forgets itself as a subject and indeed as an artistically creative subject, does the human being live with some degree of peace, security, and consistency; if he could escape for just a moment from the prison walls of this faith, it would mean the end of his "consciousness of self."[125]

The return to animality—man living through imaginary phantasms that are, at least in Nietzsche's mind, in the service of the real—is thus provided through art:

> [Art] constantly confuses the cells and the classifications of concepts by setting up new translations, metaphors, metonymies; it constantly manifests the desire to shape the given world of the waking human being in ways which are just multiform, irregular, inconsequential, incoherent, charming, and ever-new, as things are in the world of dream. Actually the waking human being is only clear about the fact that he is awake thanks to the rigid and regular web of concepts, and for that reason he sometimes comes to believe that he is dreaming if once that web of concepts is torn apart by art.[126]

One should notice the avant-gardist definition of art: art as the form of symbolic activity that is against systematic thinking and that is determined only through this task. "An aesthetic way of relating, an allusive transference, a stammering translation" is what conditions the borders of humanity, of man's consciousness of self, while at the same time presenting itself as "contradictory and impossible" but "an adequate expression of an object in the subject."[127] What is left to the visual arts, after the modern theories have separated allegories and symbols, is an animal gaze, since, as Lemm comments on Nietzsche's premises, pictorial thinking can as well be used "to generate a world of first impressions against conceptual thinking."[128] The method that Nietzsche proposes for the artists has thus as its ground a vital renunciation of humanity, a renunciation that is sought through an artistic suspension in which the thinking is left to the level of the "first metaphors." The first of the metaphors, according to Nietzsche, is always born from the process of translating the stimulation of a nerve into an "image," while the second metaphor always translates the image into a concept.[129] As Lemm argues, only the second of the two is anthropomorphic and anthropocentric, since it ensures man's self-conscious projection through linguistic representations.[130] For Nietzsche, this observation becomes essential:

> [E]verything which distinguishes [*abhebt*] human beings from animals depends on this ability [*Fähigkeit*] to displace [*verflüchtigen*] intuitive metaphors [*anschauliche Metapher*] into a scheme, in other words to dissolve a picture [*Bild*] into a concept [*Begriff*].[131]

Nietzsche's praise for unsystematic thinking systematically connects the three allied concepts: animality, the visual, and the arts. The combination produces animality as a point of return, a return obtained through the "first impressions" that give themselves visually as a "ghostly schemata."[132] These creative productions of the mind may have been, for Nietzsche, even the only way he could think one could touch the "mythical origin of things."[133] After the gods had died, animals alone represented the mythical sphere. Perhaps the return stays thoroughly mythical, but it destroys, by means of the suspense into the mythical register, the mumble of civilization.

Whatever the consequences, Nietzsche is positive that the future will prove him right. "A more advanced physiology will surely confirm that the artistic force inheres in our becoming, not only in that of the human being, but also in that of the animal: it will say that the *artistic begins* with the *organic*,"[134] he concluded.

Notes

1. Darwin, in Howard E. Gruber, *Darwin on Man: A Psychological Study of Scientific Creativity, together with Darwin's Early and Unpublished Notebooks* (Chicago: University of Chicago Press, 1981), 281.
2. Margot Norris, *Beasts of the Modern Imagination* (Baltimore, MD: Johns Hopkins University Press, 1985), 227.
3. See e.g. Kant, in Gary Steiner, *Anthropocentrism and Its Discontents: The Moral Status of Animals in the History of Western Philosophy* (Pittsburgh, PA: University of Pittsburgh Press, 2005), 170–171.
4. Jeremy Bentham, *An Introduction to the Principles of Morals and Legislation* (London: Methuen, 1982), 283n.
5. For a comprehensive account of these views in the history of philosophical thinking on the animal, see e.g. Steiner, *Anthropocentrism and Its Discontents*. Bentham's work was not published until 1789.
6. Joséphin "Sâr" Péladan, "Manifesto and Rules of the *Salon de la Rose + Croix*," in *Art in Theory, 1815–1900: An Anthology of Changing Ideas*, ed. Charles Harrison, Paul Wood, & Jason Gaiger, trans. Robert Pincus-Witten (Oxford: Blackwell Publishing, 1998), 1058.
7. See e.g. Michel Foucault, *Order of Things: An Archaeology of the Human Sciences* (London: Continuum, 2002), 245–252.
8. George Stubbs, Thomas Villiers, Earl of Clarendon & George Frederick Laurence Hyde Villiers, Earl of Clarendon, *No. I–III of a Comparative Anatomical Exposition of the Structure of the Human Body, with That of a Tiger and Common Fowl: In Thirty Tables, Two Copper-Plates to Each Table, 20 Inches by 15* (London: J. White, 1803–1806).
9. Paul L. Sawyer, "Ruskin and Tyndall: The Poetry of Matter and the Poetry of Spirit," in *Victorian Science and Victorian Values: Literary Perspectives*, ed. James Paradis & Thomas Postlewait (New Brunswick, NJ: Rutgers University Press, 1985), 217.
10. See Norris, *Beasts of the Modern Imagination*. Norris posits Darwin's works as the ground of modern biocentric thinking in the artistic field.

11. Although in many ways opposed to its views of natural philosophy, Darwin is known to have sympathized with artistic Romanticism. See e.g. Phillip Prodger, *Darwin's Camera: Art and Photography in the Theory of Evolution* (New York: Oxford University Press, 2009) and Phillip Prodger, "Ugly Disagreements: Darwin and Ruskin Discuss Sex and Beauty," in *The Art of Evolution: Darwin, Darwinisms, and Visual Culture,* ed. Barbara Larson & Fae Brauer (Hanover, NH: Dartmouth College Press, 2009).
12. See Lorraine J. Daston & Peter Galison, *Objectivity* (Cambridge, MA: Zone Books, 2010).
13. Ibid.
14. See ibid., 29–30.
15. See ibid., 30–32. Kant's idea of "objective validity" refers to "forms of sensibility," while "subjective" refers to "merely empirical sensation." Of Kant's place in the development of the concepts of objectivity and subjectivity, see ibid.
16. Ibid.
17. Coleridge cited in ibid., 30. Coleridge writes: "Now the sum of all that is merely OBJECTIVE, we will henceforth call NATURE, confining the term to its passive and material sense, as comprising all the phaenomena by which its existence is made known to us. On the other hand the sum of all that is SUBJECTIVE, we may comprehend in the name of the SELF or INTELLIGENCE. Both conceptions are in necessary antithesis."
18. An idea of animal passivity had been a cornerstone of the early modern thinking of the animal, at least since Descartes, and it had largely been unaffected by the conceptual developments that had taken place over the centuries, at least, in the major traditions.
19. As the Darwinian idea of "natural selection" is often confused with Spencerian ideas about the "survival of the fittest," which transposed the idea into the realms of sociology and ethics, it should be noted how Darwin himself grounds his idea in the principle of the manifoldness of life. Darwin defines the idea of natural selection by stating that "as many more individuals of each species are born than can possibly survive, and as, consequently, there is a frequently recurring struggle for existence, it follows that any being, if it varies however slightly in any manner profitable to itself, under the complex and sometimes varying conditions of life, will have a better chance of surviving and thus be naturally selected." Darwin cited in Barbara Larson, "Introduction," in *The Art of Evolution: Darwin, Darwinisms, and Visual Culture*, ed. Barbara Larson & Fae Brauer (Hanover, NH: Dartmouth College Press, 2009), 1; originally, Charles Darwin, *On the Origin of Species* (Calgary: Broadview Press, 2003), 97.
20. Darwin, cited in Prodger, "Ugly Disagreements," 51–52. The passage was added by Darwin to the 1869 edition of *The Origin of Species.*
21. For an interesting article on the different aesthetical premises of Darwin and Ruskin, see Prodger, "Ugly Disagreements."
22. Coleridge cited in James Paradis, "Darwin and Landscape," in *Victorian Science and Victorian Values: Literary Perspectives,* ed. James Paradis & Thomas Postlewait (New Brunswick, NJ: Rutgers University Press, 1985), 91.
23. Ibid.
24. Darwin, in Gruber, *Darwin on Man*, 46.
25. Darwin's ambiguous insistence that humans were different among the animals because of man's self-conscious faculties should nonetheless be remembered, although it loses much of its aura in the context of his view that animals are moral agents who act according to rational premises, and that morality, culture, art, language, etc. are also to be explained by the history of evolution (see e.g. Darwin, in ibid., 671; Norris, *Beasts of the Modern Imagination,* 46–49).

    It also must be remembered that Darwin did not touch upon the question of the human-animal divide in *The Origin of the Species* of 1859. Although the contestation of the boundary wasn't hard to deduce in the light of his book, it was not until the publication of the *The Descent of Man, and Selection in Relation to Sex* (in 1871) in which he explicated that, according to his views, man's life was determined by the same principles as the lives of other animals. Only in 1872, in *The Expressions of the Emotions in Man and Animals,* Darwin was ready to claim not only that we had common descents with the animals but that our expressions were also of the same nature as those they used, implying that our culture was also of an animal descent (Prodger, *Darwin's Camera,* 5–6).

26. Sara Barnes & Andrew Patrizio, "Darwin on the Threshold of the Visible: Contemporary Art and Evolution," in *The Art of Evolution: Darwin, Darwinisms, and Visual Culture*, ed. Barbara Larson & Fae Brauer (Hanover, NH: Dartmouth College Press, 2009), 290.
27. See e.g. Edmund Burke, *A Philosophical Enquiry into the Origin of our Ideas of the Sublime and Beautiful* (Oxford: Oxford University Press, 1990), 66–67.
28. Ibid., 53.
29. William Blake, *The Marriage of Heaven and Hell* (Adelaide: University of Adelaide Library), 9, accessed February 4, 2015, https://ebooks.adelaide.edu.au/b/blake/william/marriage/.
30. Burke, *A Philosophical Enquiry into the Origin of our Ideas of the Sublime and Beautiful*, 108.
31. Ibid., 77.
32. Ibid., 60–61.
33. Wordsworth, in Paradis, "Darwin and Landscape," 87.
34. Paradis, "Darwin and Landscape," 93.
35. Ibid., 94.
36. See Norris, *Beasts of the Modern Imagination*, 224.
37. Ibid.
38. Ibid.
39. Henry Fuseli, "Introduction to *The Grave*," *Art in Theory, 1648–1815: An Anthology of Changing Ideas*, ed. Charles Harrison, Paul Wood & Jason Gaiger (Oxford: Blackwell Publishing, 2000), 1009. The text is cited from an introduction that Fuseli wrote to Robert Blair's literary work *The Grave*. The 1808 edition, introduced by Fuseli, was illustrated by William Blake.
40. The birth of pictorial modernism in Manet's art (in the 1860s) has been the topic of countless art historical works. Whether or not the works consider Manet's art as signifying a rupture within the Impressionist paradigm, the theories usually argue for Manet's attempt to make his art to emphasize the visuality of the pictorial plane, the process of signification, or the ambiguous nature of facing the beholder at the expense of the secure links between the signifier and the signified. Of the different accounts of the birth of modernism in the art of Manet, see e.g. André Malraux, *The Voices of Silence*, trans. Stuart Gilbert (Princeton, NJ: Princeton University Press, 1978); T. J. Clark, *The Painting of Modern Life: Paris in the Art of Manet and His Followers* (London: Thames and Hudson, 1990); Michael Fried, *Manet's Modernism, or the Face of Painting in the 1860s* (Chicago: Chicago University Press, 1996).
41. Eugène Delacroix, *Journal, 1822–1863* (Paris: Plon, 1981), 766.
42. In connection with what I have said in the end of Part 2, the mannerization of the animal imagery in Delacroix's paintings, in contrast with Géricault's art, has often been noted (see e.g. Stephen F. Eisenman, *The Cry of Nature: Art and the Making of Animal Rights* [London: Reaktion Books, 2013], 159–160). "The scene of the fighting horses," Delacroix wrote in his *Journal* on January 29, 1832, "From the start, they stood up and fought with a fierceness which made me tremble for the sake of those men around, but the scene was really admirable for a painting" (Delacroix, *Journal*, 95).
43. Ibid.
44. Ibid.
45. See e.g. ibid., 765–768.
46. Ibid., 744.
47. Martin Jay, *Downcast Eyes: The Denigration of Vision in Twentieth-Century French Thought* (Berkeley: University of California Press, 1993), 44.
48. Ibid., 51.
49. Ibid., 29.
50. Pierre Cabanne, *Entretiens avec Marcel Duchamp* (Paris: Belfond, 1967), 74.
51. This is understandable, to give an example, under the frame adopted by Zolaesque naturalism. Zola aimed at an analysis of his subjects that was close to the scientific spirit of the times by following the positivist psychologist Claude Bernard, to "replace novels of pure imagination by novels of observation and experimentation." Zola, cited in Jay, *Downcast Eyes,* 173; originally, in Émile Zola, "The Experimental Novel," in *Documents of Modern Literary* Realism, ed. George J. Becker (Princeton, NJ: Princeton University Press, 2015), 172.

    Interestingly, Michael Fried sees that in Courbet's realism the anti-theatrical project takes on a new turn, and in some of his examples also intersects with the ambiguity inherent in animal images. According to Fried, Courbet tries to achieve the effect that the "beholder does not exist" in an extreme way when he attacks the Romantic heritage with "quasi-corporeal mergers" by claiming for a momentary illusion that the first beholder, the artist, did not exist; Courbet was trying to emphasize the process in which the artist was a part and negate the time-gap between the seeing of the image and the making of it. What makes this view so interesting is that Courbet seems in many cases to have tried to achieve an effect in which the animal, sometimes a dying animal, was used as this merger, seemingly capable of subverting the hierarchical gaze of the classical tradition. See Michael Fried, *Courbet's Realism* (Chicago: University of Chicago Press, 1990).

52. Jay, *Downcast Eyes*, 29.
53. See e.g. Delacroix, *Journal*, 95.
54. Matthew T. Simms, "The Goncourts, Gustave Planche, and Antoine-Louis Barye's *Un Jaguar dévorant un lièvre*," in *Nineteenth-Century French Studies*, vol. 38, nos. 1 & 2 (Fall–Winter 2009): 68.
55. Ibid., 75.
56. See Goncourt, cited in ibid., 71.
57. Edmond et Jules Goncourt, *Études d'art: Le Salon de 1852—La peinture à l'exposition 1855* (Paris: Librairie des Bibliophiles), 180, accessed December 18, 2016, https://archive.org/stream/gri_33125008506780#page/n3/mode/2up.
58. Barbara Larson, "Darwin's Sexual Selection and the Jealous Male," in *The Art of Evolution: Darwin, Darwinisms, and Visual Culture,* ed. Barbara Larson & Fae Brauer (Hanover, NH: Dartmouth College Press, 2009), 178; on the subject of Frémiet's work, see also Todd Gott, "Clutch of the Beast: Emmanuel Frémiet, Gorilla-Sculptor," in *Kiss of the Beast: From Paris Salon to King Kong* (Brisbane: Queensland Art Gallery, 2005), 14–57.
59. Larson, "Darwin's Sexual Selection and the Jealous Male," 179.
60. Ibid. The latter version of the work showed the woman as naked, while the former one that had produced the outrage did not.
61. Cited from Larson, "Darwin's Sexual Selection and the Jealous Male," 178; Charles Baudelaire, *Écrits sur l'art* (Paris: Le livre de poche, 1992), 314.
62. Georg Wilhelm Friedrich Hegel, *Introductory Lectures on Aesthetics*, trans. Bernard Bosanquet (London: Penguin Books, 2005), 83.
63. Gott, "Clutch of the Beast," 14–32, especially 29.
64. Baudelaire, cited in Larson, "Darwin's Sexual Selection and the Jealous Male," 178.
65. An interesting fact that also supports my reading is that Frémiet had later inscribed on the base of the former version of the sculpture that the ape was female but that it was generally understood to be a male one (Marek Zgórniak, "Frémiet's *Gorillas*: Why Do They Carry Off Women?" in *Artibus et Historiae,* vol. 27, no. 54 (2006): 224–225). For the inscription, it must be noted that it might also have been Frémiet's answer to the violent reactions of the jury and the public, since it seems that Frémiet had initially intended to picture a male gorilla (ibid.). Baudelaire, who had not seen the work at all, was of course mistaken on the fact that the pictured animal was a gorilla, not orangutan.
66. For an art-historical account of the development in question, see e.g. Martin Kemp, *Human Animal in Western Art and Science* (Chicago: University of Chicago Press, 2007).
67. See e.g. Patricia Townley Mathews, *Passionate Discontent: Creativity, Gender, and French Symbolist Art* (Chicago: University of Chicago Press, 1999).
68. A remark made by Darwin is too interesting to be left out. In a notebook, in 1868, he scribbled, "The chimaera can certainly not be classed among the 'court beauties' but should rather sit for a model to artists who desire to represent nightmares and other demons of darkness in their most hideous form" (Darwin, cited in Prodger, "Ugly Disagreements," 55).
69. See e.g. Prodger, *Darwin's Camera*, 6.
70. See Eduardo Palacio-Pérez, "Science and Belief in the Construction of the Concept of Paleolithic Religion," *Complutum* 24 (2013): 52.
71. Ernest Haeckel, *Natürliche Schöpfungs-Geschichte: Gemeinverständliche wissenschaftliche vorträge über die entwickelungslehre im allgemeinen und diejenige von Darwin, Goethe und Lamarck im besonderen* (Berlin: Druck und Verlag von Georg Reimer, 1889), accessed March 20, 2017, https://archive.org/details/natrlichesch00haec.

72. See e.g. in an artistic context Marsha Morton, "From *Monera* to *Man*: Ernst Haeckel, Darwinismus, and Nineteenth-Century German Art," in *The Art of Evolution: Darwin, Darwinisms, and Visual Culture,* ed. Barbara Larson & Fae Brauer (Hanover, NH: Dartmouth College Press, 2009), 71–72.
73. Haeckel, cited in Robert Michael Brain, "Protoplasmania: Huxley, Haeckel, and the Vibratory Organism in Late Nineteenth-Century Science and Art," in *The Art of Evolution: Darwin, Darwinisms, and Visual Culture,* ed. Barbara Larson & Fae Brauer (Hanover, NH: Dartmouth College Press, 2009), 101–103. The reference to the concept of will could be seen as a Schopenhauerian gesture. For more on Haeckel's Schopenhauerianism and its connections with the artistic culture of the time, see e.g. Morton, "From *Monera* to *Man*," 67 & 79.
74. Haeckel, in Brain, "Protoplasmania," 101.
75. Morton, "From *Monera* to *Man*," 65–66. Originally in Ernst Haeckel, "Zellseelen und Seelenzellen" (Leipzig: Alfred Kroner Verlag, 1909): 59, accessed November 17, 2016, www.zum.de/stueber/haeckel/zellseelen/index.html.
76. See e.g. Brain, "Protoplasmania," 94–101.
77. See e.g. ibid., 94–95. In one sense, the theorizations, searching for the origin, culminated in Haeckel's idea of the protoplasm; at the same time, as Thomas Huxley was to think it would serve as a mediating link between inorganic matter and organic life (ibid.).
78. See e.g. ibid., 104–117.
79. Morton, "From *Monera* to *Man*," 77.
80. See ibid., 76–82.
81. Kelley, cited in ibid., 67.
82. Stanislaw Przybyszewski, "Psychic Naturalism (the Work of Edvard Munch)." *Art in Theory, 1815–1900: An Anthology of Changing Ideas*, ed. Charles Harrison, Paul Wood, & Jason Gaiger, trans. Nicholas Walker (Oxford: Blackwell Publishing, 1998), 1045.
83. Ibid., 1046.
84. See Arthur Rimbaud, *Complete Works, Selected Letters*, trans. Wallace Fowlie (Chicago: University of Chicago Press, 2005), 369; on the French synaesthetic tradition of the late 19th century and its connections with Haeckelianism, see Pascal Rousseau, "Confusion des sens: Le débat évolutionniste sur la synésthesie dans les débuts de l'abstraction en France," *Les Cahiers du Musée national d'art modern* 74 (hiver 2000–2001): 4–33.
85. See Brain, "Protoplasmania," 107–108.
86. The modern French tradition of conceiving the concept of the gaze has been exhaustively analyzed in Jay, *Downcast Eyes*; see especially on the context of Impressionism and Post-Impressionism, pages 154–157. Anyhow, I would like to present the French views considering the connections between visuality and animality more ambiguously than as mere denigration of the visual register.
87. Henry, cited in Brain, "Protoplasmania," 104.
88. Jenny d'Héricourt, cited in ibid.
89. Félix Le Dantec, *Influences ancestrales* (Paris: Flammarion, 1920).
90. Le Dantec, cited in Brain, "Protoplasmania," 111. Translation modified.
91. See e.g. ibid., 111–112. Le Dantec wrote: "Science will perhaps lead us one day to know all phenomena of the world with just one of our senses; and that will truly be *monism*" (Le Dantec, cited in ibid., 112).
92. Though both Karl Marx and Friedrich Engels had warned against a confusion between the Darwinian struggle of life and class struggle, the relation between them was much debated in the French anarchist polemics to which part of the most important avant-garde art of the time was attached. Partly because of the strong influence of Lamarckian theories in France, the questions were quite easily appropriated into the political discourse. While the idea of the survival of the fittest was largely brushed aside in the form in which it was offered by Social Darwinism—a denial made, for example, by Kropotkin—there were even attempts to align the politics with some ideas about "Socialist Darwinism." An attempt by Armand Boucher could be named in this sense as a notable one. (See e.g. Fae Brauer, "Wild Beasts and Tame Primates: 'Le Douanier' Rousseau's Dream of Darwin's Evolution," in *The Art of Evolution: Darwin, Darwinisms, and Visual Culture,* ed. Barbara Larson & Fae Brauer [Hanover, NH: Dartmouth College Press, 2009], 214–217;

on Boucher's attempt, see Armand Boucher, *Darwinisme et socialisme* [London: Forgotten Books, 2013], accessed August 20, 2014, www.forgottenbooks.com/readbook_text/Darwinisme_et_Socialisme_1200093082/45.)

93. See Brauer, "Wild Beasts and Tame Primates," 203.
94. Max Nordau, 1895, *Degeneration* (London: William Heinemann): 142, accessed December 7, 2014, https://archive.org/details/degeneration035137mbp.
95. Ibid., 118.
96. Ibid.
97. The proverb has become famous in the 20th-century art world, since it was used by Marcel Duchamp. Duchamp said once that he had left painting behind, since he didn't want to be called "bête comme un peintre" [beastly {or stupid} like a painter] anymore. (See Marcel Duchamp, *The Writings of Marcel Duchamp*, ed. Michel Sanouillet & Elmer Peterson. [Oxford: Oxford University Press, 1973], 126.)
98. See e.g. Larson, "Darwin's Sexual Selection and the Jealous Male," 183–184. Larson writes: "Throughout his life, Munch maintained that his perpetual state of suffering and his nervous disorders were central to his creativity as an artist, not an unusual stance for a Symbolist" (ibid., 184).
99. Nietzsche's ideas on the animal have produced surprisingly few studies, taking into account the inconceivably large amount of literature handling his philosophy. As Vanessa Lemm points out, Nietzsche's ideas on the animal are central to his understanding of culture at large, and in this sense, in his philosophy: this is due to the fact that Nietzsche tries to "investigate culture [—] as a phenomenon of life." (Vanessa Lemm, *Nietzsche's Animal Philosophy: Culture, Politics, and the Animality the Human Being* [New York: Fordham University Press, 2009], 1.) One could even say that the theme has been belittled. Heidegger, to take a well-known example, explained that though the concept of the animal is a common reference point for Nietzsche, it appears only as an offshoot among his constructions. (See ibid., 157n2.) Some have also tried to maintain that Nietzsche keeps referring to the animal only for metaphorical reasons. (See e.g. Monika Langer, "The Role and the Status of the Animals in Nietzsche's Philosophy," in *Animal Others: On Ethics, Ontology and Animal Life*, ed. H. Peter Steeves [New York: State University of New York Press, 1999].) This much I hope to be able to prove inaccurate.

    It should also be mentioned, with regard to Nietzsche's thought, that different sorts of animals carry unpredictable symbolic weights, which can't be underestimated in connection to his thinking. However, as my work tries to explain how the concept of the animal worked generally within the modernist paradigm and its theory, I shall for the sake of clarity keep this theme at a distance.
100. Nordau, *Degeneration*, 415–472.
101. Friedrich Nietzsche, *The Birth of Tragedy and Other Writings*, trans. Ronald Speirs (Cambridge: Cambridge University Press, 1999), 141.
102. Ibid.
103. Ibid.
104. Ibid.
105. Nietzsche has been posited innumerable times as being both for and against Darwinism. One could say that both of the viewpoints are fully justifiable. Nietzsche himself praised Darwinism for being "the last great scientific movement" (Friedrich Nietzsche, *The Gay Science: With a Prelude in German Rhymes and an Appendix of Songs*, ed. Bernard Williams, trans. Josefine Nauckhoff & Adrian del Caro [Cambridge: Cambridge University Press], p. 218 §357), but he also wrote, characteristically, that Darwin was only a "mediocre Englishman . . . whose truths were best known by mediocre minds." (Friedrich Nietzsche, *Beyond Good and Evil: Prelude to a Philosophy of the Future*, trans. Judith Norman [Cambridge: Cambridge University Press, 2002], p. 144 §253.)
106. Nietzsche, *The Gay Science*, p. 114 §115. My emphasis. I will not discuss here the changes that Nietzsche's thought went through in the course of its development. At least with regard to the concept of the animal, the changes were not so crucial.
107. Friedrich Nietzsche, *Daybreak: Thoughts on the Prejudices of Morality*, trans. R.J. Hollingdale (Cambridge: Cambridge University Press, 1997), p. 162 §333.

108. Fernand Braudel, *A History of Civilizations*, trans. Richard Mayne (London: Penguin Books, 1993), 5. Both quotations are from Braudel's book in which they are attributed to Charles Seignobos (examples of civilization) and Eugène Cavignac (examples of culture).
109. Friedrich Nietzsche, *Digitale Kritische Gesamtausgabe Werke und Briefe* (Paris: Paolo d'Iorio), p. 75 §121, accessed March 23 2015, www.nietzschesource.org/#eKGW (1888).
110. Nietzsche, *The Gay Science*, p. 114 §115.
111. Friedrich Nietzsche, *The Antichrist, Ecce Homo, Twilight of the Idols, and Other Writings*, trans. Judith Norman (Cambridge: Cambridge University Press, 2005), p. 12 §14. Translation modified.
112. Ibid.
113. See Keith Ansell-Pearson, *Viroid Life: Perspectives on Nietzsche and the Transhuman Condition* (New York: Routledge, 1997, e.g. 102). Nietzsche himself wrote—in a paragraph named "Anti-Darwin" in the *Twilight of the Idols*—that "Darwin forgot about spirit." Nietzsche clarified further that by spirit he meant "caution, patience, cunning, disguise, great self-control, and everything involved in mimicry (which includes much of what is called virtue)." (See Nietzsche, *The Antichrist, Ecce Homo, Twilight of the Idols, and Other Writings*, p. 199 §14.)
114. Lemm, *Nietzsche's Animal Philosophy*, 2.
115. Friedrich Nietzsche, *Thus Spoke Zarathustra*, trans. Thomas Common (Paris: Feedbooks, 2010), 34.
116. Nietzsche, in Lemm, *Nietzsche's Animal Philosophy*, 18; Nietzsche, *Digitale Kritische Gesamtausgabe Werke und Briefe*, NF–1884, group 25 §428.
117. Nietzsche *The Antichrist, Ecce Homo, Twilight of the Idols, and Other Writings*, p. 201 §19.
118. Ibid., p. 201 §19.
119. Friedrich Nietzsche, *Untimely Meditations*, trans. R.J. Hollingdale (Cambridge: Cambridge University Press, 1997), p. 60 §1.
120. Nietzsche, cited in Lemm, *Nietzsche's Animal Philosophy*, 24.
121. Ibid.
122. Ibid., 114.
123. Friedrich Nietzsche, *The Will to Power*, trans. Walter Kaufmann and R.J. Hollingdale (New York: Vintage Books, 1967), p. 167 §296. Translation modified.
124. Nietzsche, cited in Lemm, *Nietzsche's Animal Philosophy*, 148; Nietzsche *Digitale Kritische Gesamtausgabe Werke und Briefe*, NF–1887, group 9 §102.
125. Nietzsche, *The Birth of Tragedy and Other Writings*, 148.
126. Ibid., 151.
127. Ibid., 148. Translation modified.
128. See Lemm, *Nietzsche's Animal Philosophy*, 116.
129. Nietzsche, *The Birth of Tragedy and Other Writings*, 144. "The stimulation of a nerve" is first translated into an image: first metaphor! The image is then imitated by a sound: second metaphor!"
130. See Lemm, *Nietzsche's Animal Philosophy*, 120.
131. Nietzsche, *The Birth of Tragedy and Other Writings*, 146.
132. Nietzsche, *The Birth of Tragedy and Other Writings*, 152. Nietzsche wrote: "No regular way leads from these intuitions into the land of the ghostly schemata and abstractions; words are not made for them; man is struck dumb when he sees them, or he will speak only in forbidden metaphors and unheard-of combinations of concepts so that, by at least demolishing and deriding the old conceptual barriers, he may do creative justice to the impression made on him by the mighty, present intuition."
133. Nietzsche, cited in Lemm, *Nietzsche's Animal Philosophy*, 120.
134. Ibid., 133; Nietzsche, *Digitale Kritische Gesamtausgabe Werke und Briefe*, NF–1872, group 19 §50.

4 The Poetic Lie

> We may, then, dispense once and for all with the idea of likeness to Nature, of correctness or incorrectness as a test, and consider only whether the emotional elements inherent in natural form are adequately discovered, unless, indeed, the emotional idea depends at any point upon likeness, or completeness of representation.
>
> —Roger Fry, *An Essay in Aesthetics*[1]

Whether or not Nietzsche's prophecies were right, the post-Darwinist fixism seems to have won the day among the avant-garde. The uniqueness of man's spirit was still mostly retained, although its divine origin was contested, and this is also one reason why contemporary theory has devalued the modern field. The picture is nevertheless more complex, as the animal often found itself at the center of the discussions over the modernist search for the new. With animality being one representative of the origin, its connectedness with the practice of art could be either affirmed or dismissed from the same foundation.

Anyhow, the reasons for emphasizing Nietzsche's work can't be thrown aside. Nietzsche's philosophy was quite influential, and it appealed forcefully to the avant-garde artists that attacked the outworn forms of Western civilization. Although Nietzsche's contribution could have implications that were a bit confused, to say the least, its ideas on animality resonated with the demands of the avant-garde in several ways. It offered three important arguments that constituted the core of the modern valuation of the animal: (1) a revolutionary value through which an animalized culture could be thrown into opposition against the oppressive nature of human civilization, (2) a sort of primitivism that promised a trespass against the intuitive origin of the arts, (3) a conceptual connection between the imaginary (or visual) register and the animal—both perceived as radically opposed to the linguistic characteristics of human representation.

These ideas framed the archetypal view of the animal in some factions of the avant-garde in the first half of the 20th century. But as we shall see, the same arguments could also be used to keep the animal away from the sphere of the arts. The constitutive indictment of the interwar period of modernism, given by Guillaume Apollinaire, that "artists are above all men who want to become inhuman,"[2] even if presented under the Nietzschean premises, could easily be turned into a statement that confirmed the exceptionality of the practice of art with regard to animal habits. Apollinaire himself affirmed that it was the artist's task to "subsist beyond the scope of nature, which strives to imprison us in that fatal order of things limiting us to the merely animal."[3] Still, the divide between the inhuman and the animal stayed equivocal, and the instability also touched upon views of the origin of art.

The Primitive Origin of Art

The primitivist currents that had been an essential part of the visual arts since the age of revolutions[4] were going through a substantial change. The reason could be found, at least in some sense, in the now extended knowledge on man's prehistory. The suppositions about prehistoric imagery varied greatly: the pictures could be used to enforce the argument that from the start man's unique nature had made it possible for him to make art, but they could also be used to defend purposes that were indebted to evolutionary theories. Be that as it may, the questions concerning the origin of the arts, and its relation to the human-animal boundary, were accented.

The idea of the return to an origin, as a way of releasing the powers of the unconscious, was not new, and it was not even new to refer to an animal origin in connection with it, but the idea developed more appeal at the end of the 19th century. Conrad Fiedler, himself an art historian, had already, in *On Judging Works of Visual Art* (1876), seen how the "peculiar consciousness of the world in development in the artist's mind" is possible to understand as some form of an "unconscious early stage of existence, like that of animals, when compared to the new consciousness of the world which man has attained."[5] Typically, in this frame of thought, the artist's art was then, according to Fiedler, related to his ability to "use nature differently," and through these means not to "pass from a lower, unconscious stage to a higher, conscious one, but rather to sacrifice the one for the sake of building up the other."[6] With his statements, Fiedler pointed to an almost canonical justification of primitivist art, while still leaving the concept of the animal somewhat untouched.

The dialectics of "sacrificing the one for the sake of building up the other" was indeed a standard formula, but its applications varied a lot. One of the most interesting explanations of the artistic nature residing in man's own origin, because of the explanation's metaphysical aura and a historical importance, was given by Wilhelm Worringer in his landmark work on modern primitivism, *Abstraction and Empathy* (1907). The work is known for its ability to make abstract forms worthy of respect in their own right (as contrasted with illusionistic art, the art of empathy), but Worringer also made known his understanding of the birth of art and its psychological implications. After he had dispensed with the connections between the natural beauty and "the condition of the work of art,"[7] he went on to speculate on the evolution that led to the first artist:

> As soon as man became a biped, and as such solely dependent upon his eyes, a slight feeling of insecurity was inevitably left behind. In the further course of his evolution, however, man freed himself from this primitive fear of extended space by habituation and intellectual reflection[8] . . . Not that primitive man sought more urgently for regularity in nature, or experienced regularity in it more intensely; just the reverse: it is because he stands so lost and spiritually helpless amidst the things of the external world, because he experiences only obscurity and caprice

> in the interconnection and flux of the phenomena of the external world, that the urge is so strong in him to divest the things of the external world of their caprice and obscurity in the world-picture and to impart to them a value of necessity and a value of regularity. To employ an audacious comparison: it is as though the instinct for the "thing in itself" were most powerful in primitive man. Increasing spiritual mastery of the outside world and habituation to it mean a blunting and dimming of this instinct.[9]

It is this instinct, then, that needs to be rediscovered. Worringer merges Hegelian and Nietzschean perspectives into his vision of the history of art, bound to return to its source: "Only after the human spirit has passed, in thousands of years of its evolution, along the whole course of rationalistic cognition, does the feeling for the 'thing in itself' re-awaken in it as the final resignation of knowledge."[10] Thus, abstraction is confirmed, with philosopher Arthur Schopenhauer's help, as means of going beyond all cognition, while the general form of alienation is understood to be an essentially human characteristic. "The losing of oneself in the contemplation of art" is, for Worringer, an evidence of the fact that "all human sensation of happiness" is due to man's "impulse of self-alienation."[11] All that really matters, then, is the inexorable return.

As noted, the ideas about an artist plunging into his own unconscious had been an important part of the discourse on the arts ever since Romanticism. But from Expressionism to Surrealism—in the era of Sigmund Freud—there appeared a new, aggressively primitivist tone supported by a weighty theoretical apparatus. It promised the artist, whose task was from day to day to dive into the darkest depths of his unconscious, that he would and should bring the treasure home: the treasure could even be a senseless abstraction, representing only the artist's own mythical origin, his primitive nature. Although the aesthetics of *Einfühlung*, still substantial in Worringer's account, kept a part of their place, the human-animal border was intensified in texts concerning art, and especially in relation to the cultural politics that led closer and closer to abstract imagery. Followers of Worringer were soon to come, one of them being Giorgio de Chirico, who, in a classic text called "Mystery and Creation" emphasized how the forms evoked by the unconscious were unrepresentable through logical linguistic means, and that they could best be invoked by beastly behavior—a reading that is known to have been directly influenced by his readings of Nietzsche and Schopenhauer.[12] This was de Chirico in 1913:

> It is most important that we should rid art of all that it has contained of *recognizable material* to date, all familiar subject-matter, all traditional ideas, all popular symbols must be banished forthwith. More important still, we must hold enormous faith in ourselves: it is essential that the revelation we receive, the conception of an image which embraces a certain thing, which has no sense in itself, which has no subject, which means *absolutely nothing* from the logical point of view, I repeat, it is essential that such a revelation or conception should speak so strongly in us, evoke such agony or joy, that we feel compelled to paint, compelled by an impulse even more urgent than the hungry desperation which drives a man to tearing at a piece of bread like a savage beast.[13]

The importance of the primitive world, its intuitive psychological quality, is affirmed when Chirico's text comes to a close:

> Perhaps the most amazing sensation passed on to us by prehistoric man is that of presentiment. It will always continue. We might consider it as an eternal proof of the irrationality of the universe. Original man must have wandered through a world full of uncanny signs. He must have trembled at each step.[14]

As already suggested, one of the reasons for the intensification of the topic was the discovery of prehistoric art, or perhaps more correctly, the theories that accompanied it. Cave art was acknowledged from the turn of the century to represent the earliest origin

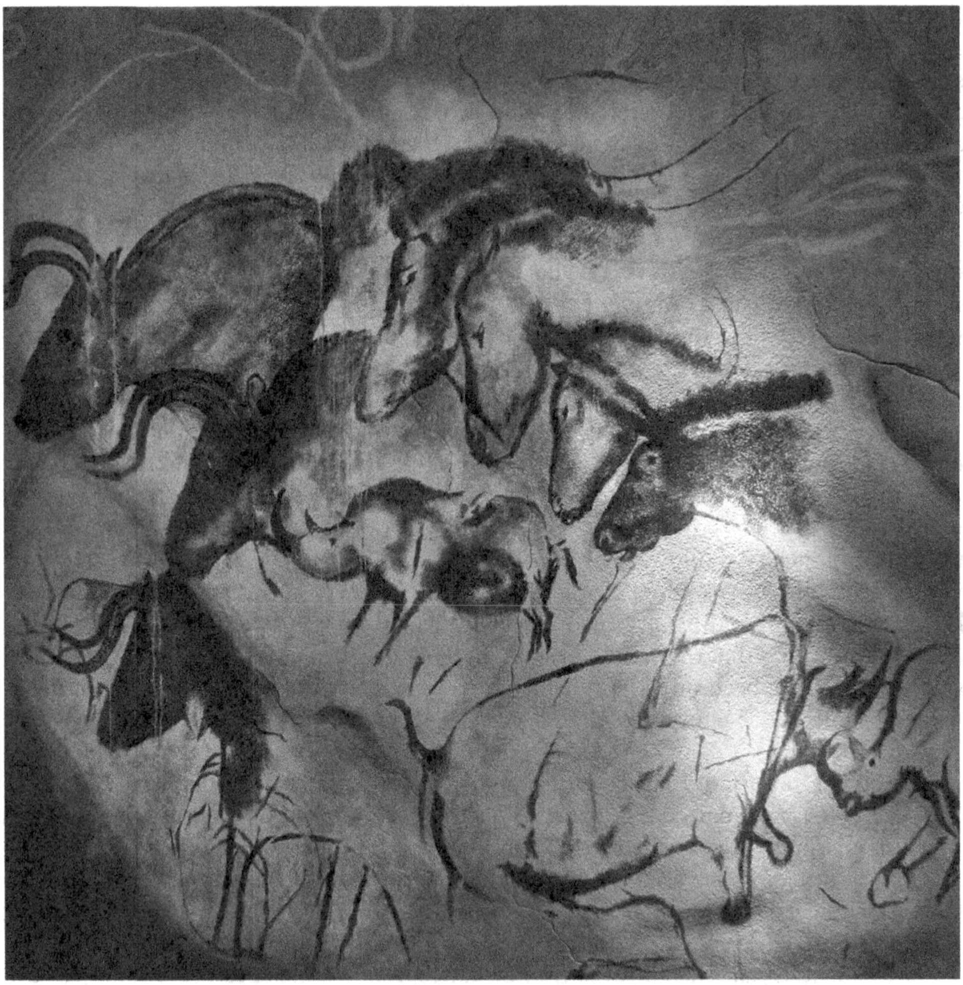

Figure 4.1 Drawings of animals in the Cave of Chauvet. c. 31,000 years old. Vallon-Pont d'Arc, Ardèche. Image by Thomas T. Available from: https://commons.wikimedia.org/wiki/File:Chauvet´s_cave_horses.jpg?uselang=fi (accessed November 7, 2016). Licensed CC-BY-SA 2.0.

of *human art*; it meant that the history of the art of painting was at least as old as the history of humanity—of *Homo sapiens*—itself. The process through which prehistoric art came to acquire this status had been going on for decades, but after the first decade of the century almost all of the leading experts in the field were beginning to affirm its right to the place.[15] After it had reached that status, it was supposed to belong to a history of practice that still continued with artistic modernity.[16]

Although the evidence of prehistoric art had received glances from the artistic community for some time—for example in Art Nouveau—it seems that prehistoric culture started to get noticed theoretically only after it had been proven that the prehistoric men may have grounded their cultures on religious beliefs.[17] This happened through work by the British evolutionist anthropologists, James Frazer, John Ferguson McLennan, and Edward Burnett Tylor.[18] Frazer's initial efforts in particular seem to have been essential. After he had introduced the idea of "sympathetic magic" (in his magnum opus, *The Golden Bough*, of 1890), it was also soon adopted by the leading experts of cave art in the disciplines of archaeology and paleontology, more closely connected with the theories of the arts.[19] Finally, Yrjö Hirn, and Salomon Reinach made explicitly known the idea that a sort of magic drove forward the origin of the visual arts,[20] with Reinach openly declaring that prehistoric man had already had "highly developed religious beliefs."[21]

However, what interests us mostly, is the rhetoric through which Paleolithic painting was presented as following the guidelines of modern aesthetics. The paintings themselves are most of all about animals—there are only a few images of the Paleolithic men themselves. The value of the idea of sympathetic magic was in this sense great, since it offered an explanation for the works. It offered an idea that the works had been scribbled in order to invoke the depicted animals—or the whole species depicted—and to improve men's luck in hunting them. This explanation, partial as it evidently was, troubled the modern minds who had, since Kant, been taught to believe in the aesthetics of indifference. How could it have been that these artists, our ancestors, were making art and practicing their religion in the name of utilitarian principles? I hope to make clear how the astonishing truth that human pictorial expression was in its early days, and for the first 20,000 years of its history, almost completely dominated by the images of animals came to be understood as a secondary characteristic in the process in which the works were integrated into the history of Western art. The fact, I assume, also tells a lot about the function of the modernist discourse at large.

The first symptoms can already be seen at work in the texts of Hirn and Reinach, both embracing the theory of sympathetic magic. Hirn, speaking of the works in connection to the origin of *our arts*, was already suspicious of calling them arts in the total sense of the word; so was Reinach. While Hirn still modestly suggested that the art of prehistoric men was "seldom free and disinterested,"[22] Reinach turned it into a problem through which we could understand the works as presenting a wholly developed religious cult, but not "art . . . as art of civilised peoples, a luxury or a pastime."[23] To integrate the works into the history of the arts, Reinach was forced, so to speak, to primitivize the primitive arts, and he did this by putting them into a Hegelian scheme in which the historical continuation between the modern and the prehistoric was understood as a process of the development of the spirit: "They show us the first steps of humanity in the path which led to the worship of animals (as in Egypt), then to that of idols in human shape (as in Greece), and finally to that of divinity as a spiritual conception."[24]

While Reinach's idea still left a place, a residue, for the image of the animal, the most influential figure in the prehistorical field in the first half of the century, Abbé Henri Breuil, did not. Although Breuil too adopted the idea of sympathetic magic, from the early texts of his career he approved only a vision of the works that maintained that the works represented the same tradition as modernist art—*400 Centuries of Cave Art*, as one of his books was named.[25] Breuil is mostly known for his systems of timing and of his methods of analyzing the works, but his more hypothetical theories were also influential, since his authority was almost indisputable in the French field, a discourse evidently leading the discussion on Paleolithic arts.

Breuil's answer to the problem confronted by Hirn and Reinach is somewhat brilliant: he indeed affirmed the idea of sympathetic magic, although he did this only in nominal terms. Breuil thought that the principle of magic had been behind the initial impulse, but that the paintings themselves, the practice in itself, were soon valued. This meant that the animal the painter had been evoking had only a secondary importance with regard to the "cult of beauty" itself.[26] Accordingly, Breuil concluded that the Paleolithic men were the first to surrender to the principle of "art for art's sake,"[27] which overcame their "commonplace utilitarianism,"[28] and appealed to their sense of immediate pleasure.[29] This art, going beyond the utilitarian needs and suppressing the initial greed that the image of the animal implicated, was thus a representative of "the first institution,"[30] a religion that marked the cornerstone of the uniqueness of our humanity, and made us special among the natural creatures. Breuil testified to this in quite radical terms:

> It is, as one of my pupils said to me in a moment of penetrating philosophical thought: Man is the only living being who takes a deep interest in what is the most useless (materially) and yet the most necessary. Meditation, so long as it leads him, by way of dreams, to cross the narrow barrier of this present life and plunge into the mysterious cosmos of what does not concern him and on which he cannot feed.[31]

For the understanding of the human need to depict animals, this left almost nothing: nothing but a worldly need to eat animal flesh. And it is not a surprise that beyond these formalist remarks and futile references to the theories of sympathetic magic encountered in "modern primitives," little attention was given to the human-animal interface.[32] More importantly, the sort of interpretation of the true origin of visual arts opened once more the possibility of dealing with the origin in a way which was considered outside the actual historicist frame and without giving much importance to the real works themselves. Under the constant need for further evidence about Paleolithic cultures and their practices, one could see the problem of the origin of the arts, once again, represented only as a cognitive fact that told us, according to an old thesis, that it was the ability to make art itself that differentiated us from animals. This is the position taken to its heights, as we have already remarked, in *The Origin of the Work of Art* by Heidegger.[33]

The intensification of the question of the essentially human need to make art, evidenced in the texts cited from Worringer and de Chirico, increased its attraction. But the premises on which the human-animal border was grounded were often supported by weak arguments. What I have tried to suggest is that the boundary and its theoretical implications had gained ground at the turn of the century because of the invocation of the conception of primitive art that facilitated the emergence of the questions concerning the boundary. The readings were thoroughly determined by the trends of modern formalism (a term here denoting pictorial formalism as well as its theoretical and philosophical counterparts, however disproportionate the two may have been). The influence of the paradigm was such that even in the context of American abstract art of the 1950s—a context mainly indifferent towards the image and the concept of the animal—a figure like Barnett Newman could explicitly think in the terms of the discourse while trying to point out art's essential function. I would like to end this section by citing Newman, since his appropriation of the dawn of art shows also the difficulties in question, the difficulties the discourse often was in danger of falling into. In his text "The First Man Was an Artist" (1947), Newman summed up his understanding of the value of paleontology in detecting the primacy of aesthetics in the minds of men:

> Original man, what does it matter who he was, giant or pygmy? What was he? That is the question for a science of paleontology that would have meaning for us today . . . Who cares who he was? What was the first man, was he a hunter, a toolmaker, a farmer, a worker, a priest, or a politician? Undoubtedly the first man was an artist . . . A science of paleontology that sets forth this proposition can be written if it builds on the postulate that the aesthetic act always precedes the social one. The totemic act of wonder in front of the tiger-ancestor came before the act of murder.[34]

The question is then, for Newman, an aesthetic one. Speech, which he understands "as an animal power"[35] is not enough, but it must be made to recognize its animal origins for the sake of the argument that tries to explain man's exceptionality, which shines through art. But this is also what makes Newman's text confused. "The human language is literature, not communication," he writes, meaning that it could not follow utilitarian needs. But how, then, could it be separated from the expressions of the animals, whose reason, for us, remains an enigma?

> The human in language is literature, not communication. Man's first cry was a song. Man's first address to a neighbor was a cry of power and solemn weakness, not a request for a drink of water. Even the animal makes a futile attempt at poetry. Ornithologists explain the cock's crow as an ecstatic outburst of his power. The moon gliding lonesome over the lake, with whom is he communicating? The dog, alone, howls at the moon. Are we to say that the first man called the sun and the stars *God* as an act of communication and only after he had finished his day's labor? The myth came before the hunt. The purpose of man's first speech was an address to the unknowable. His behavior had its origin in his artistic nature.[36]

The problem was not a new one, and it had loomed behind the classic accounts of the animal as well as those concerning the relations between animal expressions and human speech—from Plutarch to Descartes, from Montaigne to Burke. But its position had now, during the first half of the 20th century, acquired a relevance with regard to the nature of the arts. This was not only since the Paleolithic painters forced the modern mind to confront the origins of their practice in connection with the dawn of the whole of our species, but because the theory of the avant-garde had let the demand to represent the unrepresentable go so far that its practitioners had to encounter the questions: how to produce and how to depict the animal gaze?

Gaze and the Invisible

> There was stone that seemed in no way mortal, and other stone that seemed in motion, gestures that remained entirely fresh, as if they were preserved here only to be given one day to a passing child. And this vitality was not limited to the famous works, to those visible to all. The unseen, small, nameless, and seemingly superfluous works were no less filled with this deep inner force, with this rich and astonishing disquiet of life. Even the stillness, where there was stillness, consisted of hundreds of motive moments held in equilibrium. There were small figures, especially animals, moving, stretching, or crouching, and even when a bird sat still, one knew very well that it was a bird, for the sky issued forth and surrounded it, a breadth was apparent in the smallest folds of its wings, which could spread to astonishing size.
>
> And the same thing was true of the animals that stood and sat on the cathedrals, or crouched beneath the consoles, bent and bowed and too inert to bear weight. There were dogs and squirrels, sparrows and lizards, turtles, rats, and snakes. At least one of every kind.
>
> These creatures appeared to have been captured out in the forests and on the paths, as if the strain of living among shoots, flowers, and leaves of stone had transformed them slowly into what they were now and would always remain. But there were also animals that were born into this world of stone, without any memory of another existence. They had always been entirely at home in this upright, towering, precipitous world. Skeletons arched up among these fanatically lean creatures. Their mouths opened wide with cries of the deaf, for the nearby bells had destroyed their hearing. Some crouched like birds upon the balustrades, as if they were passing through and simply wanted to rest for a few centuries, staring down at the growing city. Others, descended from dogs, thrust horizontally from the edge of the spouting into the air, prepared to spit water from swollen maws. All these creatures had adapted and changed, but they lost none of their vitality in the process. To the contrary, they lived more vigorously and more violently, they lived eternally the fervent and impetuous life of the time that had given rise to them.
>
> Seeing this picture, one sensed that these creatures had not resulted from a whim, or from a merely playful attempt to find new, unusual forms. They were born of necessity.[37]

The writer of the text is not Eugène Viollet-le-Duc, nor John Ruskin, nor William Morris. Neither is it written to eulogize the art of the cathedrals, though it refers to them. Its subject is the art of Auguste Rodin. And the writer behind the text is Rainer Maria Rilke, who also worked as Rodin's secretary for a while.[38] The animal imagery here evoked is not easy to comprehend with relation to Rodin's works (the text indeed tries to praise Rodin); although there are animals in Rodin's works, he certainly wasn't known as an *animalier*. The stakes are a bit different.

From the first texts of his career to the last, Rilke made explicit that there was a similarity between the gaze of the animal and the gaze of the modern artist, and that in depictions of animal life one could find something that was essential for the human arts. Animality, then, was a quality that emanated from an artwork, and it could be made to speak of the relations that the artist had with the world. This relation was specific: it was beyond the utilitarian usage of the things of the world. There was one difference, a difference that made up the question of method, but which was also used as an authorization of the arts. In Rilke's romantic mind, animals lived in a "night without things,"[39] while men could only acquire the same relation to the world by listening for "the sound of things."[40] In a famous paragraph of the 8th *Duino Elegy*, Rilke explained the grounds of his thinking:

> All eyes, the creatures of the World look out
> into the open. But our human eyes,
> as if turned right around and glaring in,
> encircle them; prohibiting their passing.
> What lies outside, their faces plainly show us.
> Yet we compel even our youngest; force
> each child always to stare behind, at what's
> already manifest, and not to see
> that openness which lies so deep within
> the gaze of animals[41]

The open, reflected in the animal's eyes, is evidently only a consequence of Rilke's affirmation of the ancient argument that only humans possess self-consciousness, and that only men are able to contemplate their own death. However, Rilke says that "the eternity in which animals move"[42] can be glimpsed, but only for a moment, if the animal gaze is somehow reflected in the things we create for eternity. Even then it perhaps stays, for us, as only a mere simulacrum of animal life:

> Death leaves beasts free.
> Only we foreknow it. Animals
> keep death behind, and before them, God.
> And when a beast passes, it passes in
> eternity, as river run[43]

Still, a few passages later Rilke shows what the sight of an animal can do to man:

> Our gaze is ever turned towards Creation;
> we know only the surface of that glass,
> its clouded image, by our selves obscured.
> And yet, sometimes a silent animal
> looks up at us and silently looks through us.
> We call it Fate to be in opposition.
> Nothing but that. Forever opposite.[44]

There is then a double demand relative to the animal: images should depict animals, but that is not enough, since the equity with animal life should also shine through the work and gestures made by the artist. The latter is what Rilke tries to show about Rodin's art with his all too numerous references to the animals. Rodin's work is, for Rilke, an animal cathedral, but not because it inhabits the world with images of animals made of stone—although they cause no harm—but, following the idea that there is an isomorphism between the artist's mind and the way the audience sees the picture, "since the directness, strength, and warmth of pure animal life"[45] emanates from Rodin's rapid sketches of his models, and since "the rest [and] the greatness would have to come [to the work] only when everything was finished, just as animals come down to drink when the night is full and there are no longer strange things in the forest."[46] It is the "joy of dogs"[47] that the artist's life, for Rilke, embodies.

And of what, indeed, is the joy of dogs made? The essence of things must be explored in depth if it is to be understood correctly; only then can it briefly appeal to human understanding. Cézanne's art was another target of Rilke's formalist practices of animalizing the artist's work. Rilke wrote in 1907 of Cézanne that the artist's work showed that everything had "become an affair that's settled among the colors themselves." The naïve formalist remark was soon sustained by an animal metaphor. Rilke continued to fantasize about Cézanne's animal condition by making an interesting juxtaposition that suited his premises: "just as in the mouth of a dog various secretions will gather in anticipation at the approach of various things—consenting ones for drawing out nutrients, and correcting ones to neutralize poisons: in the same way various intensifications and dilutions take place in the core of every color, helping it to survive contact with others."[48] There are dozens of these delicious examples in Rilke's texts, which testify to his conviction that the animal condition could be glimpsed through modern practices of art aligned with formalism. The problems arrived when the idealist principles had to be applied to artistic practices. How, then, to achieve the animalizing effect with certainty, if the conventional image of the animal was not enough?

If not for Rilke, then at least this posited an actual problem for some artists. One of these artists was Franz Marc, who, in the first decades of the century, made himself known for his formalist depictions of animals. Interestingly, the aesthetics of *Der Blaue Reiter* group could lead to quite different outputs with regard to the animal, as the example of Wassily Kandinsky will soon show us. In any case, the instruction that these artists were to offer connected the image of the animal in the center of the history of the birth of abstract art. The animal, of course, was not the only path that led to abstraction: as Marcella Lista has noted, the artists of the group tried to validate their early attempts at abstract art with such different points of reference as animals, crafts, oriental ornaments, and music, while finally attaching themselves to the effort of creating a new man and a new society.[49]

However, Marc took Nietzsche's challenge seriously, and made explicit that his aim was to "animalize art" (*Animalisiering der Kunst*).[50] This was to be achieved by the means of creating "the animal painting" (*Tierbild*).[51] *Tierbild*, as in Rilke's double demand, did not mean only that the new art should be filled with animal images; Marc himself thought that artists should also "animalize the artistic sensation" (*eine Animalisierung des Kunstempfindens*).[52] Marc explained, for a start, that this sort of painting epitomized "a sensibility to the organic rhythm of all things, a pantheistic feeling into the quivering and flowing of the blood of nature,"[53] and confessed that he thought that some of the modernists had already achieved the effect. "With van Gogh or Signac," Marc testified, "everything has become animal: the air, the very boat that rests on the water, and above all, painting itself."[54] Despite the fact that Rilke had his reservations about the art made under the Expressionist label, the combination did not go unnoticed for him; on the contrary, Rilke was overwhelmed by the nature of Marc's solution. In 1916, after visiting Marc's retrospective in Munich, Rilke wrote that "almost no one foresaw that it would be so significant, finally once more an *œuvre*, an achievement in the work and an attained unity of life, and how blessed, unconditioned, pure,"[55] still later stating that the exhibition had been an "event" (*Ereignis*)[56]—a charged concept in Rilke's context—for him.

And Marc's solution? It is not a surprise that Marc's choice leaned towards a formal language that was at the same time understood as the new way for the Western art of painting to overcome the restrictions of the categories of time and space: Cubism. Cubist aesthetics also defined for the moment, for Marc, a possible way to consolidate the musical rhythms of nature and timeless images: the merging of the two could open an animal gaze by confusing the two categories whose understanding was the support of man's consciousness of himself. This was Marc's way to set the task in a focal text, in which he asked "How does a horse see the world?":

> Is there a more mysterious idea for an artist than to imagine how nature is reflected in the eyes of an animal? How does a horse see the world, how does an eagle, a doe, or a dog? It is a poverty-stricken convention to place animals into landscapes as seen by men; instead we should contemplate the soul of the animal to divine its way of sight . . . The Cubists were the first to paint space not as a subject: they "said something about" space, they rendered the predicate of the subject . . . Who is able to paint the existence of a dog as Picasso paints the existence of a cubic shape? (In the thematic style of musicians).[57]

As one could assume, a couple of years later Rilke was on the same trail. Some glances towards the inner life of the animal were demanded from the artist. This "in-seeing" complemented Rilke's former ideas: the outer form of an animal was now becoming secondary. Only such seeing could answer the demands of modern avant-garde: one should not only "see through" but be able to "see in"—there is perhaps no better way of distinguishing with four words between the visual aims of the Renaissance and high modernity. This "seeing in" was a momentary pause, a sort of a leap. What arts could do, in the end, was somehow to sublimate the leap and leave it open for the beholder. This leap should go all the way down, to the place "where dog starts to be a dog," but it should be made with care, Rilke reminds us: the poetic leap towards animal sphere, towards a sphere where all familiar forms and all common denominators may disappear, is also a leap of madness that any project of becoming one's own other always keeps within. This is Rilke at his best, in 1914:

> I love in-seeing [*Einsehn*]. Can you think with me how marvellous it is, for instance, to see into a dog while walking by, in-seeing (I don't mean seeing-through [*Durchschauen*], which is, in contrast, only a kind of human gymnastics and where one immediately comes out again on the other side of the dog, only, as it were, regarding it as a window into the human lying behind it, not this)—letting oneself into the dog, exactly into its centre, to the place where it starts being a dog, there, where God, as it were, would have positioned himself for a moment, when the dog was complete, in order to oversee its first embarrassments and incidents and to affirm that it was good, that nothing was lacking, that one could not make it better. For a while one can endure being in the centre of the dog, one must only watch out and be sure to leap out in due time, before its environment [*Umwelt*] wholly encloses one, because otherwise one would simply remain the dog in the dog and be lost for anything else.[58]

The last sentence may also explain why the concept of the animal kept its eligibility even for the darker and revolutionary side of the avant-garde, whether or not the artists

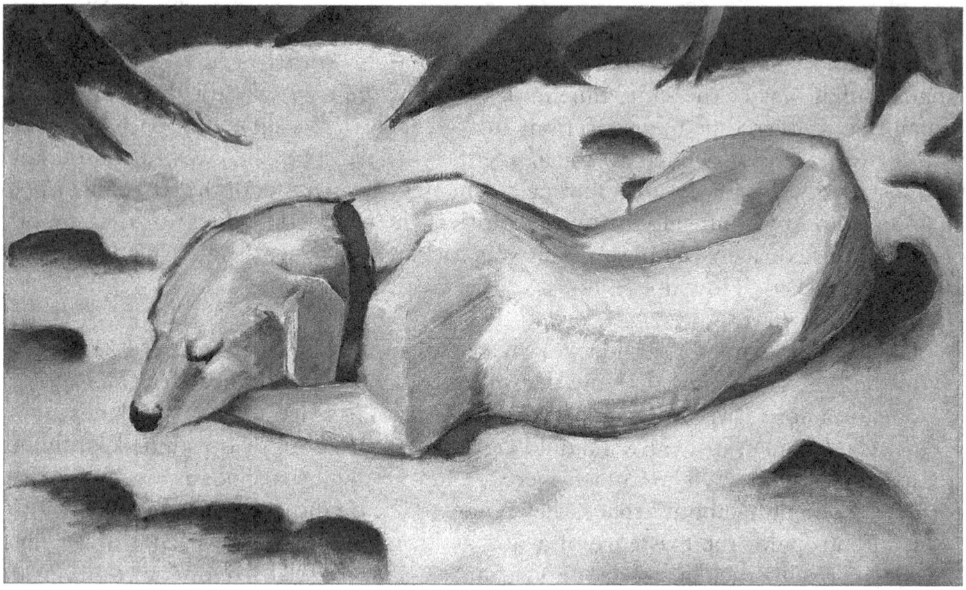

Figure 4.2 Franz Marc, *Dog Lying in the Snow*. 1911. Oil on canvas, 105 x 62.5 cm. Städelschen Kunstinstitut, Frankfurt am Main. Available from: https://commons.wikimedia.org/wiki/File:Franz_Marc_-_Dog_Lying_in_the_Snow_-_Google_Art_Project.jpg?uselang=fi (accessed November 7, 2016).

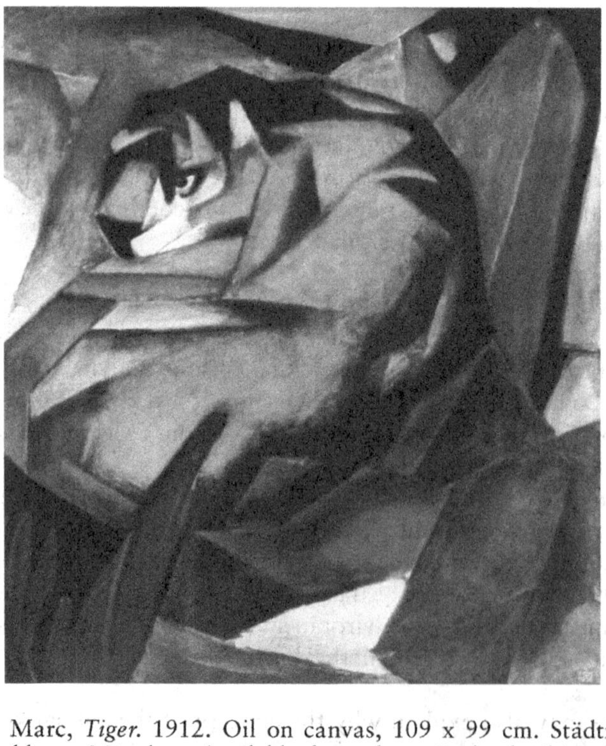

Figure 4.3 Franz Marc, *Tiger*. 1912. Oil on canvas, 109 x 99 cm. Städtische Galerie im Lenbachhaus, München. Available from: https://upload.wikimedia.org/wikipedia/commons/6/61/Franz_Marc_027.jpg (accessed November 7, 2016).

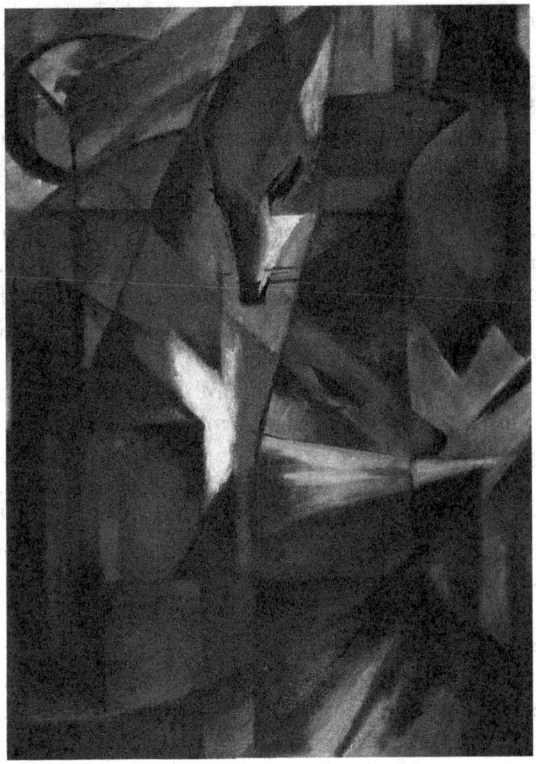

Figure 4.4 Franz Marc, *The Foxes*. 1913. Oil on canvas, 88.0 x 66.0 cm. Museum Kunstpalast, Düsseldorf. Available from: https://upload.wikimedia.org/wikipedia/commons/0/0c/Franz_Marc-The_Foxes_-Google_Art_Project.jpg?uselang=fi (accessed November 7, 2016).

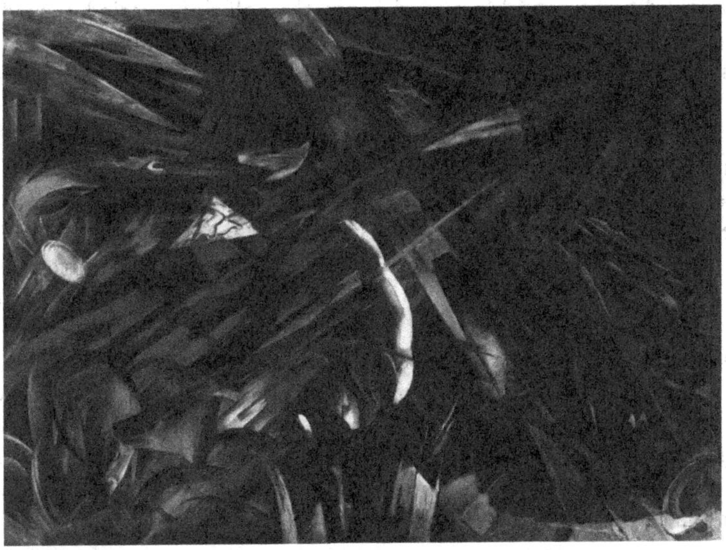

Figure 4.5 Franz Marc, *The Fate of the Animals*. 1913. Oil on canvas, 196 x 266 cm. Kunstmuseum Basel. Available from: https://commons.wikimedia.org/wiki/File:Franz_Marc-The_fate_of_the_animals-1913.jpg?uselang=fi (accessed November 7, 2016).

preferred to proceed through the means offered by the dreams of abstract art. To leave the topic for a moment, what mattered both for Marc and Rilke was the inner structure of the animal gaze, which was posited against a superficial art clinging to outer appearances. Actually, it meant that a curious combination took place: the view demanded a dialectics between the individual expressionist demands and the supposedly universal forms that could adequately reflect the conditions of animal experience by the means of human art. This task was Marc's essential problem during the years before the war.

Although Rilke and Marc have become the most renowned of the searchers for the animal spirit in their time, and deservedly so, they were not alone, and different conceptual developments were continuously at play. The artistic discourse, often influenced by Nietzschean currents, took already canonized modern artists one at a time as examples of the animal gaze that could be made to appear through the formalist structure of modern art: Munch, van Gogh, Signac—the legacy was easily named. An Expressionist critic, Theodor Däubler, even thought that "tracking animal categories" was the central quest for all Expressionist artists. Däubler made clear with his reference to Impressionist art—presented in contrast with Expressionism—that it was only through projections of inner life that the animal could break through:

> Am I not an animal? A reality! The cycle A and O says: you can be one. The animal breaks through in Munch, as a full expression of his wholly unbroken essence. The Impressionists did not know quite what to do with it. But now, once again, we are on the track of animal categories—*Tieberstimmheiten*. Above all, with monumentality: every animal an undeniable grasping at life. A consequential self-determination. The return to the animal through art is our decision in favor of Expressionism.[59]

Whether or not the animal broke through in Munch, the dialectics between the inner and the outer did posit a problem, and again the solutions varied according to the context. The Nietzschean demand—that one should overcome humanity in favor of the animal by means of human culture—seemed oddly, when transposed into the practice of modern art, to turn into a strict formalism, for lack of a better decision. In other words, the overcoming of the boundaries of humanity was conceived less to be a question of content than of form.

In this frame of thought, the concept of the animal could easily again turn into a mere metaphoric device. The Expressionist problem of the inner and the outer facilitated the development: if the formal demand for natural expression was the only thing that mattered, if the linguistic representations by themselves were not as important as the inner necessity of making them, then only the principle of making art, whatever that was thought to be, really mattered. Surely, animal life could stand as a metonymy of the practice of art—a conception of it could even be used as a law that determined what was art and what was not—but if art was a practice of a certain style brought out by a particular kind of mental process, the output was often that what was left to the animal was but a metaphoric value. August Macke could give a poetic example of the problems of the discourse:

> Man expresses his life in forms. Each form of art is an expression of his inner life. The exterior of the form of art is its interior.

> Each genuine form of art emerges from a living correlation of man to the real substance of the forms of nature, the forms of art. The scent of a flower, the joyful leaping of a dog, a dancer, the donning of jewelry, a temple, a painting, a style, the life of a nation, of an era.
>
> The flower opens at sunrise. Seeing his prey, the panther crouches, and as a result of seeing it, his strength grows. And the tension of his strength shows in the length of his leap. The form of art, its style, is a result of tension.[60]

The organic metaphor, the idea that a work of art should be done according to organic measures, or in harmony with inner life, was not a new one. Neither was the idea about a tension between man and forms that Macke suggested. The organic metaphor was far from being new, and it had even been hegemonic from time to time in the history of arts (for example, during the Renaissance). However, there was a new rigidity in the way in which it was now posited, and it was a part of the dialectics of the inner and the outer in which the animal sometimes had a role, sometimes didn't. Perhaps it was man's duty to create forms that seemed to be as alive as animals, and perhaps, this was the extent to which the idea of the animalized image mattered. A renowned quote from Odilon Redon could illustrate the idea: his organicist ideal of the arts called for making "unreal creatures live humanly, by putting, as much as possible, the logic of the visible at the service of the invisible."[61] Although Redon was writing beyond the tradition discussed so far in this section, his thesis might serve as a summary of the pursuit of picturing the idealized otherness connected to animality in the tradition that had now, most notably, transposed the principles of Nietzsche's animal philosophy into an avant-gardist art determined by the formalist demands of the context. At least for a moment, the animal had seemed to be heading in the same direction as the politics leading to abstract art.

The World of Abstraction and the Revolution of the Beasts

In several ways, the discussions of the invisible prolonged the influence of Goethe's theories, especially as this was the current among the fashionable neo-Platonic theories in which the interplay between the phenomena of nature and inner sense was conceived.[62] Goethe had believed that the phenomena of nature had also a true *inner* appearance that could be embodied through an artwork, and conversely, that the great works of art were produced according to true and natural laws that were hidden behind outer appearances.[63] From the theosophical currents of the early decades of the 20th century to Wassily Kandinsky's *Point and Line to Plane* (1926) and Paul Klee's *Wege des Naturstudiums* (1923), such Goethean dialectics were embraced.[64] The two works mentioned, one could say, represent a point in synthesizing the organicist ideal of art and the visual art of abstraction, an idea epistemologically indebted to the tradition of German Romanticism and especially to Goethe's ideas of the ideal artist and his relation to nature.[65] But the modern ideal of the autonomy of the work of art, its self-importance, posited in this sense a question concerning its relations to the animal: if the animal's spirit was not considered free in the human sense, how could one think of an animal being comparable in any way to a gaze, posited in the work of art, which on the contrary was evidence of the free and creative mind humans had? Both Kandinsky and Klee tried to find the laws behind natural imagery, and the quest

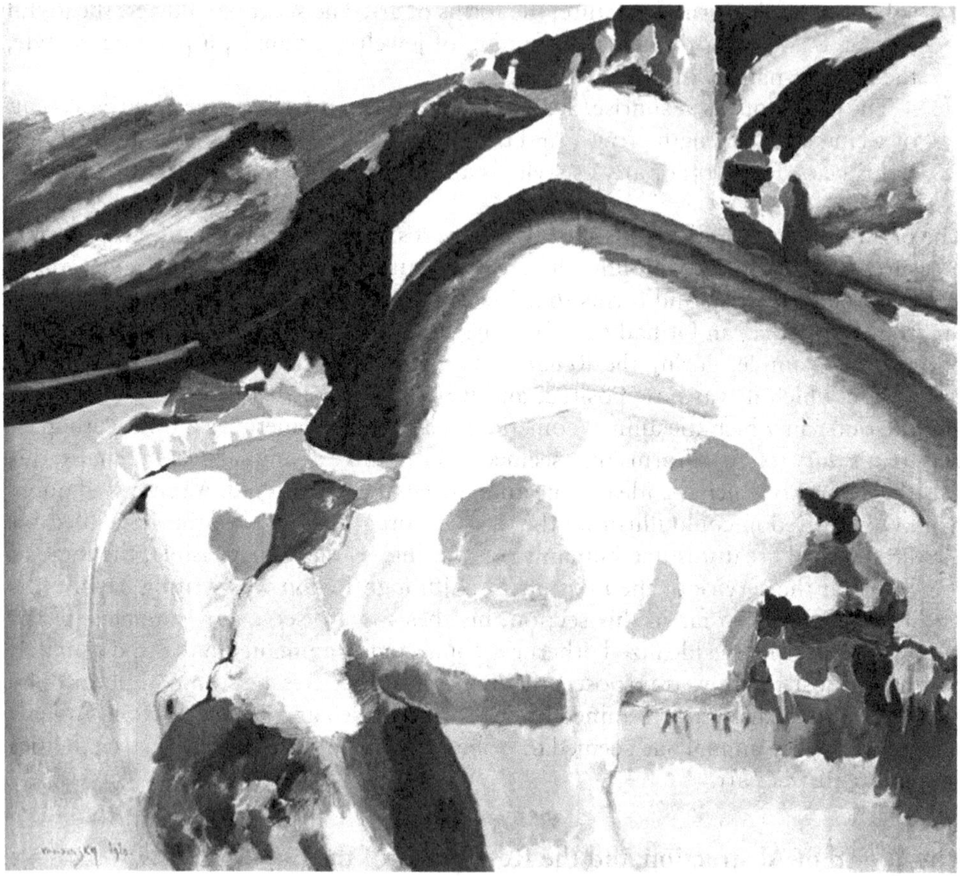

Figure 4.6 Wassily Kandinsky, *The Cow*. 1910. Oil on canvas, 95.5 x 105 cm. Städtische Galerie im Lenbachhaus, München. Available from: https://upload.wikimedia.org/wikipedia/commons/b/b5/Vassily_Kandinsky%2C_1910_-_The_Cow.jpg (accessed November 30, 2016).

also authorized their abstract works, but by following Goethe's example, both only attempted to "humanize"[66] their natural objects. It seemed that the laws of nature were hidden, and only artistic and spiritual human intuition could get hold of them.

The animal had played a somewhat minor role in the early works of Kandinsky. The artist himself had produced a series of paintings depicting animal life, while he was on his way towards abstract art, concurrently with Marc's works already discussed. However, Kandinsky soon took a more radical turn, when his "improvisations" led him definitively to the world of abstract art, and from 1912 onwards he explicitly denied the role that animality could have played in this development. Thus, in *Concerning the Spiritual in Art* (1912) he emphatically noted how the creative impulse was inherent only in human nature, not belonging to the animal, while at the same time he contrasted modern art with the classicist paradigm, of which the latter was only a "mere aping."[67]

The remark should not be read only as metaphor, since Kandinsky was later to announce that "love of nature" and "indefinite stirrings of the urge to create" were "fundamentally different" impulses, and in the first page of his book, he depicted how "externally the monkey completely resembles a human being," although "[the monkey] will sit holding a book in front of his nose, and turn over the pages with a thoughtful aspect, but his actions have no real meaning for him"[68]—one should only wonder why Kandinsky thought it so important to tell all this to his readers at the beginning of his essay! Nevertheless, the oppositions were supposed to be understood as crucial to his theory of art.[69]

As influential as Kandinsky's view of abstraction may have been, the animal maintained its attraction in the different conceptions of abstract art that were soon to come; all did not accept Kandinsky's assessment of classical art as "apings." Filippo Tommaso Marinetti's *Manifesto of Futurism* had already made extensive use of animal metaphors, putting them in the service of his revolution,[70] but it was perhaps his Russian counterpart who acknowledged the revolutionary potential of the animal more acutely. Kasimir Malevich's texts, written before the October Revolution, accented the revolutionary potential of the concept of the animal by stressing the untamedness and authenticity of animal nature, setting it against humanity and classical culture. In contrast with Kandinsky's judgments, Malevich's rhetoric—through which the concept of the animal was exploited quite rashly—thus introduced the tradition of classicism in terms of what was worst in humanity, and what was totally alienated from our animal side. The revolutionary rhetoric, contrasting the authentic life of the wild beasts with oppressive human civilization, was of course old, but it seemed to acquire new, thoroughly formalist aspects in Malevich's mind.

In *From Cubism and Futurism to Suprematism: The New Realism in Painting* (1915–1916), written during the years when the question of the nature of the oppressed had become intensified, Malevich defended his own position as an artist following the ideological guidelines of the revolution and, in the pictorial sense, those of abstraction, which he thought possible to consolidate. Spiritual revolution was coming, and it had to make itself felt in the visual arts: "The gallop of a horse can be depicted with a pencil of one colour, but it is impossible to depict the movement of red, green or blue masses with a pencil," Malevich wrote, demanding that painters, "if they wish to be pure painters," should "abandon subject and objects."[71] The contemptuous tone towards the image of an animal—a careful choice: the horse, the noblest of the animals in the classical sense—did not, however, lead Malevich to abandon the value of the animal. On the contrary, the vital impulse that the animal could be seen to represent, was shown in its connections to the new visual language that Malevich was creating. The blame went from Rome to the Russian Empire, both of which had supported the classical rules of art. The text is Malevich at his utmost, defending the animal issue to the last breath and condemning all exploitation of nature, from the upkeep of zoological gardens to the practice of fishing, by favoring a new anti-classicist art:

> Is this not torture?
> Setting the soul in marble and then mocking the living.
> But your pride is an artist who knows how to torture.
> You put birds in a cage also for pleasure.
> And for the sake of knowledge you keep animals in zoological gardens.

> I am fortunate to have broken out of that torture-chamber of the Inquisition which is academicism.
> . . . I have released all the birds from the eternal cage, and opened the gates to the animals in zoological gardens.
> May they tear to pieces and devour the remains of your art.
> And may the freed bear bathe his body in the ice of the frozen north and not languish in the aquarium of boiled water in the academic garden.
> . . . The instrument of torture is idealism and the demands of aesthetic feeling.
> The idealisation of the form of man is the mortification of much living sinew.
> Aestheticism is the garbage of intuitive feeling.
> You want to see pieces of living nature on the hooks of your walls.
> Just as Nero admired the torn bodies of people and animals from the zoological garden.
> I say to all: reject love, reject aestheticism, reject the trunk of wisdom, for in the new culture your wisdom is laughable and insignificant.
> I have united the knots of wisdom and set free the consciousness of colour!
> Remove from yourself quickly the hardened skin of centuries, so that you may catch us up the more easily.
> I have overcome the impossible and formed gulfs with my breathing.
> You are in the nets of horizon, like fish!
> We, Suprematists, throw open the way to you.
> Hurry!
> For tomorrow you will not recognize us.[72]

It evidently did not matter that Malevich's politics of abstract art had now torn to pieces even the image of the ferocious animal itself by picturing the revolution only through abstract black-and-white imagery. And indeed, after some time, when the Soviet revolution had taken place, one would not recognize the Suprematists. Only four years later, Malevich already thought it suitable to exclaim, from within the post-Revolutionary state, that "now only he—man—as a centre can turn nature into another new image, which will be nothing less than man himself: a completed step on the eternal path."[73]

After the First World War, the latter attitude was the one predominantly taken by the defenders of abstract art. Somewhat paradoxically, this was the case even when the artists made explicit that their abstractions had their origin in natural forms, as was the situation with Kandinsky and Klee in the 1920s and 1930s.[74] While form and the unity of an artwork came always first—a lesson possibly deduced from the Goethean parallelism between living organisms and artworks created according to the hidden laws of nature—this left some theorists in a position from which they had to have recourse to some kind of predisposed "natural" order of forms that were hence cut off from any of their relations to animality—a position that could finally be maintained only by enhancing the ontological dualism living within human nature, between man's human and animal aspects. In this sense, abstraction was the pure art that opposed the animal urges living inside us. Piet Mondrian's "Plastic Art and Pure Plastic Art" of 1937 is almost canonical in its formulations with this in mind:

> The laws which in the culture of art have become more and more determinate are *the great hidden laws of nature which art establishes in its own fashion*. It is

> necessary to stress the fact that these laws are more or less hidden behind the superficial aspect of nature. Abstract art is therefore opposed to a natural representation of things. But it *is not opposed to nature* as is generally thought. It is opposed to the raw primitive animal nature of man, but it is one with true human nature. It is opposed to the conventional laws created during the culture of the particular form but it is one with the laws of the culture of pure relationships.[75]

The artistic revolution kept its aura, but in a utopian dimension, in which the unity of the composition was the highest aim. There was no place for the animal: the postwar avant-gardes were often possessed by the utopian task of, as Jean Clair has called it, "the making of the new man"[76]—a task to which Kandinsky, Mondrian, and Malevich all connected themselves[77]—much more than with anything that should identify with the oppressed as Malevich had done before the war and for which the animal rhetoric would have been useful. And in a similar way, the conception of the animal was mostly forgotten in American formalism, in spite of some attempts to revive the discussions. In this sense, even the tiny concessions that Kandinsky and Klee were ready to give to natural and animal forms seemed to echo some sort of avant-gardist conservatism in the climate of the 1930s.

When the formalist conception of the arts was carried to such heights, the role of the animal could be forgotten. This was the tradition to which the organicist and the naturalist metaphor of the arts easily surrendered. As in Heidegger's mind, artwork could in this sense even acquire a state of life that surpassed that possessed by nonhuman animals. The epistemological crisis seemed to have come to its conclusion, allowing the arts to get rid of the image of the animal altogether—and with it, of all the representative functions of art—letting the theorists at the same time continue glancing condescendingly towards the natural sciences.[78] This was also the condition in which Alfred H. Barr found abstract art and introduced it to the American community of art lovers by guiding them to see abstractism, meaning "the term most frequently used to describe the more extreme effects of the impulse away from 'nature'."[79] It is telling about the dual nature of the discourse that Mark Rothko could equally think in 1947 that his ideal artist should consider artistic shapes as "organisms with volition and a passion for self-assertion."[80]

Nobody perhaps had felt the fatality of the consequences—the development connecting the artistic revolution with the tendency to dispense with the representations of nature, while still believing that art possessed a utopian dimension within it—as deeply as Franz Marc himself. Marc, writing quite touchingly from the front in 1915, testified how the lure of abstract purity and idealist beauty had swallowed up the world. Marc's remark gives a chilling account of the process that needs no addition:

> Another instinct led me from animals to abstractions, which roused me even more, it brought me to that second sight, timeless in the Indian sense in which the living feelings shine in their purity. Very early I found people to be "ugly": animals seemed more beautiful, more pure. But then I discovered in them, too, so much that was ugly and unfeeling—and instinctively by an inner compulsion, my presentation became more schematic or more abstract. Trees, flowers, the earth, all showed me every year more and more of their deformity and their repulsiveness—until now, suddenly I have become fully conscious of nature's ugliness and impurity. Perhaps

> it is our European point-of-view that makes the world look poisoned and distorted. This is why I dream of a new Europe.[81]

Dream, Debauchery, Myth

The revolutionary potential of the animal did not go unnoticed by the avant-garde, and it was exploited in several ways, some of them new. The primitivist characteristic of animal art—now assuredly in continuation with the Western heritage because of the interpretations made of the cave paintings found in the Franco-Cantabrian Region—gave the image of the animal a certain revolutionary aura. Modern art, which promised to go beyond the restrictions imposed by the very conception of art, often resorted to primitivist strategies. From naïvism to *art brut*, from Fauvism (literally, "beastly art") to Outsider Art, and from Cubism's devotion to exotic fetishes to Surrealist eccentricities, the image of the animal carried some primitivist value that was used as a tool to go beyond the all-too-rational, all-too-human basis of Western art. Mostly, the discourses were still organized around the modern conception of the primitive, a quality that was valued either positively (paralleling it with romanticized ideas of the origin and authenticity), or negatively (leading to a diagnosis that all such works made in the modern context were signs of a pathological or a degenerate mind).

Another of the traditions was embraced by early Surrealism, as well as by the different conceptions of "anti-art." It derives from the Nietzschean-Rilkean views, but is also indebted to Kantian aesthetics of indifference. Basically, the tradition maintains that animal life and the process that leads to good avant-garde art are both essentially oppositional to the world of work. Curiously, the view found itself often in combination with Marxist politics.

The heyday of this tradition was the years between the two world wars. Within the boundaries of the discourse, the value of the animal is determined by a certain conception of work: what the world of work consists of, according to the tradition, is a mediated relation to the world. Work is mediation; to the subject of work, work represents its object until the work gets done. The animal's relation to the world and the experience of art—what Surrealists called "poetry," connecting themselves with certain sides of Romanticism—was on the contrary thought to be immediate and immanent. This was of course the way in which Rilke and Marc had already considered animals as beings that lived "in the open." Among the lineage of the avant-garde, leading from Dada to Fluxus via Surrealism, the movements dispensing with the durability of the work of art, it was the most influential. Anti-art was supposed to be anti-work, non-productive in itself—descriptively, George Maciunas claimed in the 1960s that "a flight of a butterfly or movements of microbes" were "anti-art."[82]

As in Rilke's thinking, according to which the poet's language is essentially different from the language of everyday life,[83] the animal is among the traditions seen as an other for the human mind, another life that man can assert only through poetic means. In other words, animality is considered a pure state of otherness which is nonetheless laid down as a human possibility that can be fulfilled through artistic processes. Sometimes, the twofold relation is even used as a tool for explaining man's own animality, although only for the sake of affirming the underlying dualism that is by its nature ontological. The frequent usage of the concept of the animal in anti-artistic discourse—that is, in a discourse that demanded that art should overcome its own specificities, or that it should merge with life—can be comprehended in the light of the ambiguity that it posits between the animal and the human: within this context, art can always turn into its opposite, anti-art, in the same way as man can let his identity be affirmed by a glimpse of his own animality. It is only because of these obscure patterns that Michel Tapié could observe, reminding us of Apollinaire's proclamations, that "since Nietzsche and Dada art had seemed the most inhuman of adventures, from beginning to end."[84]

As suggested, the most important carrier of the tradition was the Surrealist movement. The Surrealists often tried to make incursions into a conceptual field based on the ambiguity of the human-animal interface, while attempting to submerge art into life (or, at least, to represent the conditions for such submerging). But as the movement itself was subject to much ideological turmoil, they also made their historical mark on the discussion, and the destiny of the concept of the animal was often marked by this movement's internal conflicts.

The first decade of Surrealism was colored by its characteristic interpretations of Freudianism and Marxism, and as such they also determined the role of the concept of the animal. The movement attacked the bourgeois understanding of the concept of work and stressed the meaning of the powers of the unconscious in the task of liberating man from the constraints of civilization. While the ideas about animality were often based on a combination of Freudian symbolism and Nietzschean idealism, the Surrealist variation was often distinguished from the earlier currents, since it accented the meaning of the moment of seeing an animal, or, the moment of seeing an image of an animal in a dream-like state.

A fine example of the value attached to the animal can be found in Paul Éluard's little book of poems titled *Les animaux et leurs hommes, les hommes et leurs animaux* ["Animals and their men, men and their animals," 1920]. The book was composed of magnificent little pre-Surrealist poems, all dedicated to different animal species. The most interesting of the poems was, regarding the concept of the animal at large, a poem dedicated to man, called "Homme utile," since it showed how easily the Rilkean ideal of animal life was understood in an early Surrealist context as a sort of a concrete inability to work—to be useless to the world and to the machinery of bourgeois economics. The whole poem is structured around the idea that animal world can archetypally represent the side of life that becomes manifested as anti-utilitarian and consumptive—whatever the context in which the animal appears ("desert," "fields," "forests," "seas" . . .) What is new in Éluard's account, among the early 20th-century writers, is the pre-Surrealist vision of how an image of an animal is able to stimulate man, giving "free rein"[85] to one's luxury moments of dreaming. This is Éluard's account:

Homme utile

You can't work any more. Dream!
With open eyes, with open hands
In the desert
In the desert which plays
With the animal—the useless.

After order, after disorder
In the plains, in the empty forests,

In the heavy and clear seas,
An animal passes—and your dream
Has really become a dream of repose.[86]

Perhaps the most interesting aspect of the history of Surrealism of the 1920s in this regard is the fading of the Nietzschean-Rilkean idealism, or at least, its transformation. The progression evidently coincides with "the return to reason" of the artistic avant-garde at large, but while the other art movements seem to have generally dispensed with the concept of the animal, the Surrealists, on the contrary, made it one of their central points of reference. Much of this was indeed indebted to their keen but unorthodox readings of the works of Sigmund Freud.

The value of the animal, in the early 1920s, is for the Surrealists mostly in the otherness that the animal world represents. The economical view, represented by Éluard's poem, shows the other side of the picture; the other is, as suggested, the Freudian world of desires and dreams relative to Surrealist ideas on *psychic automatism*.[87] Freudian texts, evidently, have their own animal iconography. However, as the 1920s was coming towards its end, the darker shades of Surrealist Freudianism started to come to the fore: the writings of Georges Bataille, the works of Salvador Dalí, Alberto Giacometti, and André Masson—three artists who were soon excommunicated from the group, partly because of the violent character of their works—and the Surrealist enthusiasm for the books of the Marquis de Sade all reveal this darkness. They all testified to a newly intensified interest in the animal: the psyche had been finally opened up to Freudian monsters, and thus, the dream of unoccupied animality started to fade away.

From the start, the Surrealist interest of Freud is centered on sexuality, and the Surrealist conception of the animal was no exception. Freud himself had remarked that when our ancestors had developed into human beings by adopting a bipedal position, they had also delivered man from the world of olfactory experiences that had been crucial to man's animal sexuality.[88] Little by little, a human culture had developed, a culture that tried to protect men and their societies against natural impulses.[89] The theme was saluted by the Surrealists when Freud made it explicit in his *Civilization and Its Discontents* (1930). In the book, Freud defined the word "civilization" [*Kultur*] as describing the whole sum of the cultural achievements and regulations which distinguished our lives from those of our animal ancestors, and thought that civilization primarily served two purposes: it protected men against nature and adjusted men's mutual relations.[90] Comprehensibly, this increased the dangerous allure of the animal in the eyes of the young Surrealists. Freud's idea of the civilization as "an apparatus whose most elementary purpose is to consolidate and transmit the defensive illusion of man's uniqueness, to perpetuate this foundational rejection of an animal outside,"[91] as Nicholas Ray has put it, may only have sounded like an exhortation to submerge oneself into the animal world. The value of the assumption was grounded on the underlying idea that civilizations suppress the *internal* animal enemy of man by making it *external*.[92] "The urge for freedom, therefore," Freud had concluded, "is directed against particular forms and demands of civilization or against civilization altogether."[93] This was taken as a guideline by the second generation of the Surrealists when they attached themselves to the task of showing the nature of man from the perspective of animal impulses.

The animal opening is then, when transposed to the Freudian discourse on sexuality, understood as a space where animal innocence—its "pure" sexuality—is made to affirm the perverse nature of human eroticism, marked by the strict social rules binding it.[94] As many of the Surrealists understood, this view could be used to generate an effective image of the monsters inherent in the Freudian unconscious of the modern man. However, the deepening of the vision towards the new Freudian terrains during the 1920s—*Beyond the Pleasure Principle* was written in 1920, and the ideas presented in it little by little reached French thinkers—soon opened more complicated visions on animal sexuality, and the Surrealist references soon tended to strain away from the domain of innocence. Freud had opened the discourse to the ambiguous terrain between Eros and Thanatos, and the Surrealists followed.[95] The double nature of the derogation brought about by the negative use of the concept of the animal—the idea that animals are "uncivilized," and that man's animal nature is assembled from his socially undesirable characteristics—was about to be overturned.[96]

What seems to be the most important aspect of the turn is that when the earlier conceptions of the animal, which nonetheless emphasized the sexual undercurrents of animal imagery, connected themselves with an idea of the innocence of the animal world, the violent nature of animality was later presented as unsuppressed and explicit. One of the key points was the fascination that led to a Surrealist insectomania in the early 1930s. When the early years had been dominated, as usual in the arts and human imagination, by mammals, Surrealist imagery was now full of insects and related animals. Memorable examples can be found in Luis Buñuel's and Salvador Dalí's cinematic masterpieces *Un chien andalou* (1929) and *L'âge d'or* (1930): the first presenting a famous image of ants coming out of a man's hand; the latter opening with an anti-clerical statement, in which an image depicting clergymen is juxtaposed with a scene of fighting scorpions. Much of the attraction towards insects was indeed indebted, or at least accelerated, by an entomologist whose works Buñuel—once a student of entomology—admired.[97]

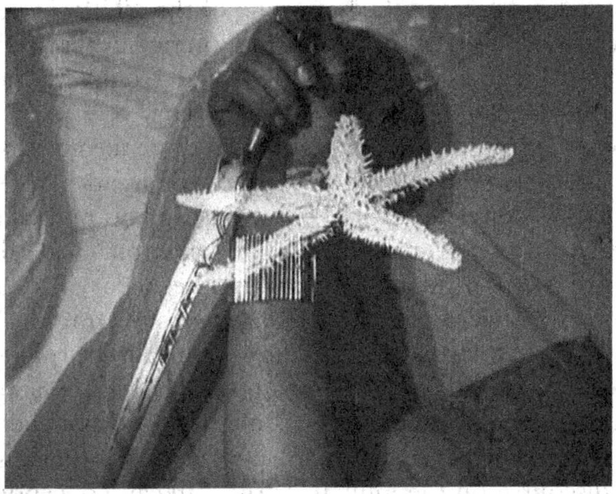

Figure 4.7 Man Ray, still from *L'Étoile de mer*. 1928. 35 mm film. Directed by Man Ray. From: *Avant Garde—Experimental Cinema of the 1920s & 1930s*. DVD. New York: Kino Video, 2005.

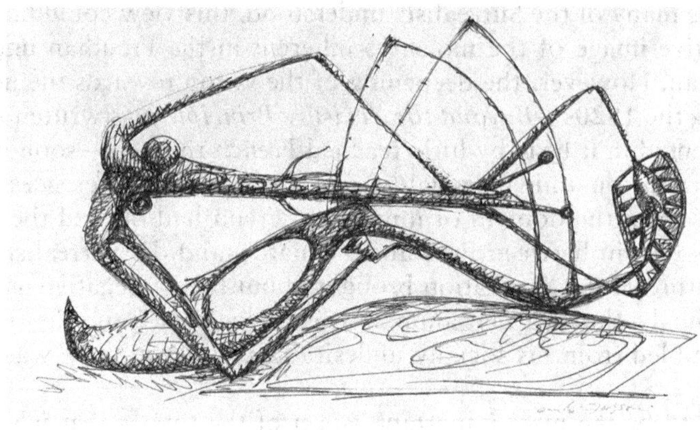

Figure 4.8 André Masson, *Praying Mantis*. c. 1942. Ink on paper, 34.9 x 58.4 cm. Museum of Modern Art (MoMA), New York. Gift of Philip Johnson. Acc. no.: 595.1970 © Digital image, The Museum of Modern Art, New York/Scala, Florence, 2016.

The object of Buñuel's admiration was Jean-Henri Fabre, nowadays best known for his book *Souvenirs entomologiques*. One of Fabre's minute descriptions of the habits of insects particularly attracted the Surrealists: the one depicting a female mantis having sex with seven of her husbands after biting off their heads. "And the masculine remnant," Fabre cried out, "firmly anchored, continued its duty!"[98] Many of the Surrealist writers were so in love with this true-life horror story, made up of a macabre connection between eroticism and death, a connection that many of them heralded as a key to understanding life, that they even kept mantises as their pets, and many more of them—including Dalí, Giacometti, Victor Brauner, and Masson—used some kind of a mantis-suggesting form as a central part of their iconographies of the era. The mantis also found its place, in the beginning of the 1930s, in perhaps one of the most famous Surrealist methods of seeing and interpreting artworks: Dalí's paranoiac-critical account of Jean-François Millet's *Angelus* suggested that the pose of the woman in the picture was the one of a mantis preparing to mate.[99]

The mantis also had its theoretical day. A peak in these theorizations of the era was produced by Roger Caillois, who in the beginning of the decade wrote texts that would become minor classics in "French theory"—texts nowadays particularly remembered because of Jacques Lacan's referring to them in his paper on "The Mirror Stage."[100] In "The Praying Mantis" (1934) and in "Mimicry and Legendary Psychasthenia" (1935) Caillois brought forward an idea of an animal impulse, similar to the Freudian "death-drive," which according to the young author now seemed to be ruling the habits of the insects and transcending the vulgar Darwinian notion of self-preservation.[101]

To be sure, Caillois's texts are not only a manifestation of the role that animal had in the works of the Surrealists in the early 1930s, but they also signify a kind of rupture in the discourse on the animal. If the Rilkean conception of the animal looking on to the open had progressed, in the minds of the Surrealists, towards an understanding of the violent unconscious animal forces through new prioritizations within the Freudian theory, it was now, because of the same development, on the brink of losing its value on the dualist scheme of things. One can see symptoms of this everywhere in Surrealist imagery, which, towards 1935, had become populated with hybrid forms combining human and animal images, particularly specializing in insects and birds. The possibility of mirroring human violence with this sort of animal imagery is hinted at already in the former text by Caillois, in which he declared that the anthropomorphic qualities of the mantis have an unusual ability to touch the deeper layers of the human unconscious.[102] Caillois compared the mantis to a bat and its several human attributes—of which he mentions "the presence of real hands, pectoral breasts, periodic menstrual flow, and a free, dangling penis"[103]—and concluded by announcing that the praying mantis manifests man's "inclination to be interested in, or even identify with, anything whose external configuration suggests his own body."[104] One is led to think of the Nietzschean fantasy of animality or the imaginary identification discussed in Lacan's theory.

The most exemplary case of the development of the Surrealist animal imagery can, without a question, be located in the *œuvre* of Max Ernst. From the haunting little images of his early works like *The Master's Bedroom, It's Worth Spending a Night There* (c. 1920) and the *Two Children Threatened by a Nightingale* (1924) on, Ernst gradually developed a peculiar Freudian relationship to birds—he readily admitted having connections with his childhood memories of (Ernst speaking of himself in the third person) "a dangerous confusion between birds and humans which became fixed in his mind and asserted itself in his drawings and paintings."[105] Later in the 1920s, Ernst even imagined for himself a bird-formed superego called "Loplop, Bird Superior." When the 1930s came, Ernst had already developed into a fantastic teratographer, his great work in this sense being the graphic novel *Une semaine de bonté*, first published in 1934, in which the reader is forced to confront numerous sadistic and torturous habits of the half-animal, half-men figures. Margot Norris hits the point when she comments on Ernst's lions that they "do not stalk and devour their prey: they kidnap it in carriages and hold it for ransom."[106]

In about 1934, the same year that Surrealism finally broke away from its announced alliance with the French Communist Party (PCF), all things seem to have started to change again: the imagery started to get milder, and the sadistic themes were often put aside. Animals were usually still made to represent the unconscious forces as, to take an example, Gavin Parkinson has noted from Max Ernst's sophisticated 1936 *Joie de vivre,* seeing its "semi-concealed insects and small creatures pointing to the instinctual forces of desire allowing free rein outside of the organizing structures of the conscious mind."[107] Surrealist animal imagery now became more philosophical, suggestive and abstract, and less exploitative and anti-rationalistic. Fine examples can be found—to mention a few—in the metaphysical postwar paintings of Remedios Varo, from the ever-developing animal iconography of Ernst, from the enigma paintings by René Magritte, and from the wonderful series of pictures by Masson, called *The Anatomy of My Universe* (1943). The question of the animal became philosophical, as the questions asked by the Surrealists were also becoming involved in the polemics determined by the philosophies of difference and those originating in Existentialism.

Although the playfulness of the early animal imagery was not totally forgotten, the characteristically Surrealist way was now to orient animal imagery towards a fashionable theorization of myths, a discussion also carried out via the visual arts: founding an iconography of the so-called Surrealist myths, or making the images sound political

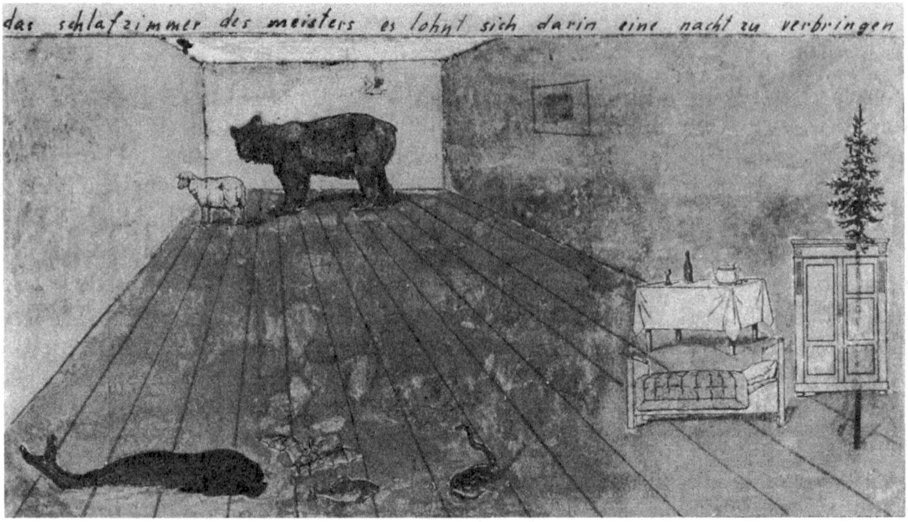

Figure 4.9 Max Ernst, *The Master's Bedroom, It's Worth Spending a Night There*. 1920. Collage, gouache, and pencil on paper, 16.3 x 22 cm. Private collection. Available from: https://dome.mit.edu/handle/1721.3/20013 (accessed November 8, 2017).

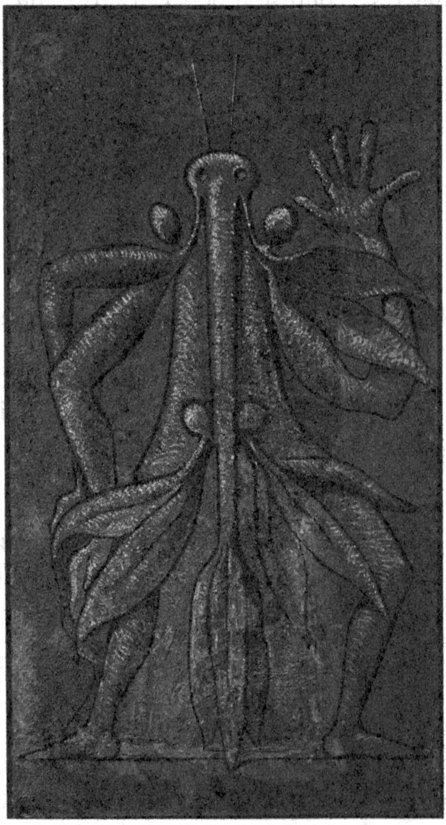

Figure 4.10 Max Ernst, *Figure humaine*. 1931. Oil on plaster on wood, 183 x 100 cm. Moderna museet, Stockholm. © Moderna museet 2016.

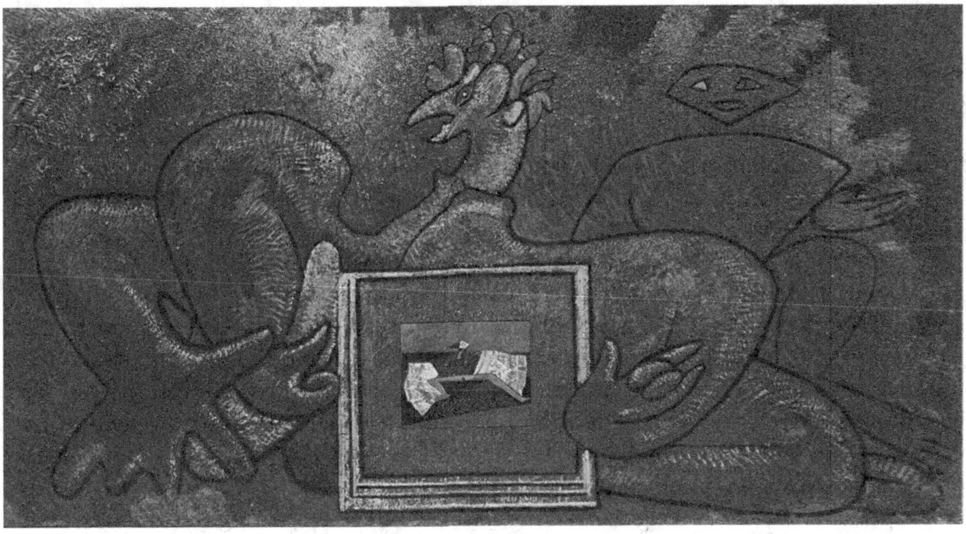

Figure 4.11 Max Ernst, *Loplop Introduces Loplop*. 1930. Oil and various materials on wood, 100 x 180 cm. Private collection. © Kuvasto 2016.

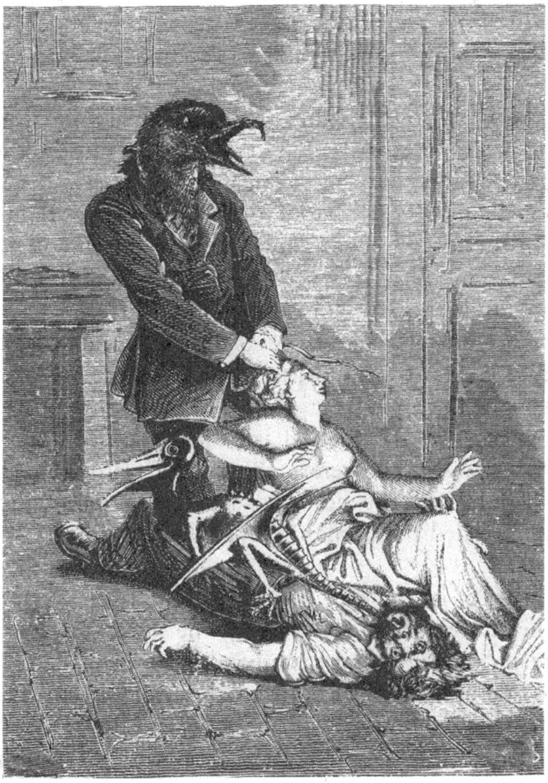

Figure 4.12 Max Ernst, Excerpt from *Une semaine de bonté*. 1934. Etching. © Kuvasto 2016. From: Max Ernst: *Une semaine de bonté: A Surrealistic Novel in Collage*. New York: Dover Publications, 1976, p. 140.

by resisting the totalitarian regime of the myth-making of the era by using the image of the unsociable animal.[108] The practice of myth-making reached its peak in the *Paris Exhibition of Surrealist Art of 1947*. The exhibition was mostly covered by a so-called "Labyrinth of Initiations," which was made up of 12 "altars," each one dedicated, according to subscription, to "a being, a category of beings, or an object susceptible of being endowed with mythical life."[109]

The turning point was again also manifested in a text by Caillois, "Function of Myth," published in 1936 and revised in 1938.[110] In the text, Caillois makes animal life an origin of the myth, and conversely, the myth a condensation of the animal forces. After a period of stressing the violent nature of animality, Caillois moved far away from the understanding of animality as the antithesis of the industrious humanity mocked by the early Surrealists. Although he had detached himself from the movement, there is no question about whether he was influenced the most by the Surrealist atmosphere of the time. Now, the political question was about the structuring of the myths, and of the possibilities of resisting the vicious circle of utility by putting the human animal in charge of its own myth.[111] To illustrate this, Caillois chose a citation taken from Victor Henry:

> Myth certainly predates man: for an organism endowed with some degree of consciousness, any apperception of an external fact is a potential myth. For the brain of a higher animal, the universe is expressed as a series of myths, that is, of instantaneous representations that vanish as soon as they occur.[112]

The heaviest proof, however, of the value of the mythical animal life among the Surrealists, was given in 1947 by Georges Bataille, a writer mostly known for his erotic novels and his extra-academic career as a radical philosopher. This was done in a work published posthumously in 1974 but based on the conferences Bataille gave at the end of the 1940s on the history of religions.[113] In *Théorie de la religion*, Bataille argued that the human fantasy of animal life, "poetic lie of animality" as he called it, is the basis for all human culture, and especially for religion and the arts.

Like Nietzsche before him, Bataille insisted on the visual nature of the fantasy of animality. What mattered was the opposition that it erected against rational premises; the value of the image of the animal, or an image reminding us of the animal condition, consisted in the momentary poetic flash that it could induce.[114] Bataille's idea of the fantasy may have needed to renew the dualist frame, but it also went all the way to its border: it emphasized the moment when the border no longer mattered. The fantasy of animal life needed the human-animal dualism, but only to overcome it in "immanence."[115]

The dream of the sort of immanence consisted of an overdetermination of the concept of the origin: the animal origin was not just the starting point for human history, but the living, ungraspable origin that kept existing as the ground for our artistic and religious phantasms. Bataille emphasized that the idea of animal life, as a life submerged into an immanence, should not be accepted without reservations, and that the idea itself was even "questionable,"[116] but that the conception of animal life, understood as being "like water in water,"[117] was in any case the presupposition that kept the fantasy alive, and with the fantasy, the whole cultural edifice that it supported by conversely claiming that humanity meant an ability to transcend man's own self. The poetic moments—keeping up the animal life that continued in us "tenderly, secretly, and painfully"[118]—were the ones that dispensed with the utilitarian relation with the world, with the subject-object relationship with which we enslaved ourselves and others, and from which the momentary glimpses of animal immanence could give us rest. This was art's task in the avant-gardist frame that Bataille wanted to introduce.[119] The structure of the fantasy—isomorphic with the opposition between the profane world of work and the sphere of the sacred[120]—was, according to Bataille, respectively based on our mutual fantasy of ourselves as totally self-conscious, leaving the outside for animality: the theory represented us as subjects of representation, animals as image. Bataille sang, to a Rilkean tune, of our fantasy of visualizing the animal landscape, one without objects, *impossible* to conceive.[121]

Notes

1. Roger Fry, "An Essay in Aesthetics," in *Art in Theory, 1900–2000: An Anthology of Changing Ideas*, ed. Charles Harrison & Paul Wood (Oxford: Blackwell Publishing, 2003), 82. Despite the ambiguity of the statement, one should add that Fry's views, when considering animals in the context of his aesthetics, were quite clear: "The whole of animal life, and a great part of human life, is made up of these instinctive reactions to sensible objects, and their accompanying emotions. But man has the peculiar faculty of calling up again in his mind the echo of past experiences of this kind, of going over it again, "in imagination," as we say. He has therefore, the possibility of a double life; one the actual life, the other the imaginative life." (Ibid., 76.)
2. Guillaume Apollinaire, "Pure Painting," in *The Modern Tradition: Backgrounds of Modern Literature,* ed. Richard Ellmann & Charles Feidelson, Jr. (New York: Oxford University Press, 1965), 114.
3. Ibid. "Painfully they search for traces of inhumanity, traces which are to be found nowhere in nature," Apollinaire writes. A few remarks later, Apollinaire indeed refers to Nietzsche, but in the same text oddly calls for a modern "anti-Greek spirit"—whatever the implications, the spirit was supposed to provide the reason for the artist to raise himself above mere descriptions of nature.
4. For a general view of the topic, E. H. Gombrich, *The Preference for the Primitive: Episodes in the History of Western Taste and Art* (London: Phaidon, 2002). Gombrich's argument is that primitivism was born again in 18th-century historicism, being thus a modern phenomenon but having its roots in antiquity.
5. Conrad Fiedler, *On Judging Works of Visual Art*, trans. Henry Schaefer-Simmern & Fulmer Mood (Berkeley: University of California Press, 1978), 49–50.
6. Ibid., 50–51.
7. Wilhelm Worringer, *Abstraction and Empathy: A Contribution to the Psychology of Style*, trans. Michael Bullock (Chicago: Ivan R. Dee, 1997), 3. It must be admitted that Worringer does not dispense with the concept of natural beauty without reservations: "despite the fact that in the course of evolution [natural beauty] seems to have become a valuable element in the work of art, and to some extent indeed positively identical with it." Ibid.
8. Ibid., 16.
9. Ibid., 17–18.
10. Ibid., 18.
11. Ibid., 24–25.
12. See Riccardo Dottori, "The Metaphysical Parable in Giorgio de Chirico's Painting," in *Metafisica*, no. 5–6 (2006): 203. Chirico: "Nietzsche and Schopenhauer were the first to teach me the non-sense of life and how such non-sense could be transmuted in art." (Ibid.)
13. Giorgio de Chirico, "Mystery and Creation," in *Art in Theory, 1900–2000: An Anthology of Changing Ideas*, ed. Charles Harrison & Paul Wood (Oxford: Blackwell Publishing, 2003), 58.

14. Ibid.
15. A landmark in this sense was Émile Cartailhac, "La grotte d'Altamira, Espagne. Mea culpa d'un sceptique," in *L'Anthropologie 13* (1902) in which he, as an authority in the field, confessed his mistake of accusing the cave of Altamira of being a fake. For a more extended account on the developing perception of art's beginning in prehistory, see for example Eduardo Palacio-Pérez, "The Origins of the Concept of 'Palaeolithic Art': Theoretical Roots of an Idea," in *Journal of Archaeological Method and Theory*, vol. 20 (December 2013).
16. For my article on the topic and the development of the modern appropriation of cave paintings, see Roni Grén, "Luolamaalaukset taiteena—uskonnollisen tulkinnan variaatioita Salomon Reinachilla, Abbé Breuililla ja Georges Bataillella," [Cave Paintings as Art—Variations of Religious Interpretation in Salomon Reinach, Abbé Breuil, and Georges Bataille] in *Tahiti—taidehistoria tieteenä* 1/2015, accessed June 25, 2015, http://tahiti.fi/01-2015/tieteelliset-artikkelit/luolamaalaukset-taiteena—uskonnollisen-tulkinnan-variaatioita-salomon-reinachilla-abbe-breuililla-ja-georges-bataillella/
17. See e.g. Eduardo Palacio-Pérez, "Cave Art and the Theory of Art: The Origins of the Religious Interpretation of Palaeolithic Graphic Expression," in *Oxford Journal of Archaeology*, vol. 29 (2010).
18. See ibid. Tylor introduced the concept of animism in 1871, McLennan the concept of totemism in 1869, and Frazer the principle of sympathetic magic in 1890.
19. See e.g. ibid., 3–6. See also, James George Frazer, *The Golden Bough: A Study in Magic and Religion* (Oxford: Oxford University Press, 2009).
20. Yrjö Hirn most notably in his book *The Origins of Art: A Psychological and Sociological Inquiry* (London: Macmillan, 1900); Reinach most famously in Salomon Reinach, *Cults, Myths and Religions*, trans. Elizabeth Frost (London: David Nutt, 1912), 124–137.
21. Salomon Reinach, "Paul Girop et Emile Massenat, les stations de l'âge du Renne dans les vallées de la Vézère et de la Corrèze," in *Revue Archeologique*, 3ème série (1899): 478.
22. Hirn, *The Origins of Art*, 12.
23. Reinach, *Cults, Myths and Religions*, 135.
24. Salomon Reinach, *Apollo: histoire générale des arts plastiques* (Paris: Librairie Hachette, 1904), 8. Hegel had suggested, in *Aesthetics*, a historiographical model according to which the history of the spirit traveled through symbolical, classical, and romantic phases. In the symbolical, the animals were used, as "abstract attributes" (Georg Wilhelm Friedrich Hegel, *Introductory Lectures on Aesthetics*, trans. Bernard Bosanquet, [London: Penguin Books, 2004], 83), while in the last of the phases "the Idea . . . has to reveal itself in the medium of spirit and feelings as perfected in itself" (ibid., 88). In connection with our topic, see also Part 1.
25. Already in 1906, Breuil (with Cartailhac) had commented on the nature of the Paleolithic works, suggesting a connection with the modern arts. They had argued that the works could be seen as "a collective manifestation, governed by traditionally established rules, in which perpetuity was perhaps ensured by a caste of people who transmitted and zealously guarded the rules of the art and the notion of its magic value." Henri Breuil & Émile Cartailhac, *La Caverne d'Altamira à Santillane près Santander (Espagne)* (Monaco: Imprimerie de Monaco, 1906), 135; Eduardo Palacio-Pérez, "Science and Belief in the Construction of the Concept of Paleolithic Religion," *Complutum*, vol. 24 (2013): 57.
26. Henri Breuil, *Four Hundred Centuries of Cave Art*, trans. Mary E. Boyle (New York: Hacker Art Books, 1979), 11.
27. Henri Breuil, "Paleolithic Art (art mobilier)," manuscript archived in *Fonds Breuil*, Br 48.8, Muséum national d'histoire naturelle, Paris, 12–13. "That if art for art's sake had not been born, magical or religious art would never have existed; but if magical and religious ideas had not allowed the introduction of art for its own sake in the most pressing daily anxieties of human existence (such as the pursuit of food, etc.), the latter having but a feeble hold on the ordinary occupations of daily life, ran the risk of remaining in the embryonic stage."
28. Breuil, *Four Hundred Centuries of Cave Art*, 11.
29. Unlike Reinach, who thought that the works themselves were not made for the sake of an immediate pleasure but to evoke the animals (Reinach, *Cults, Myths and Religions*, 135), Breuil argued that the cult was based on visual gestures that gave immediate pleasure to the

persons watching them, and thus, by means of transference, helped them to overcome their worldly desires and impulses—among them, the need to eat the animals (see Breuil, *Four Hundred Centuries of Cave Art*, 11).

30. Ibid., 22.
31. Ibid., 10. Translation modified.
32. Few attempts were made, however. One of the most well-known was the attempt of Max Raphael, who tried to read the works through totemism. Raphael read the pictures of the animals by imagining them to contain symbolic values that belonged to the totemic system. See Max Raphael, *Prehistoric Cave Paintings*, trans. Norbert Guterman (New York: Pantheon Books, 1945). Another attempt to give the image of the animal more visibility was that made by anthropologist Hans Richter, who suggested a shamanistic function—a theory that has received much attention in the last decades because of the researches made by Jean Clottes and David Lewis-Williams. See Horst Kirchner, "Ein Beitrag zur Urgeschichte der Schamanismus," in *Anthropos*. t. 47 (1952), 244–286; for the later developments of the thesis, see Jean Clottes & David Lewis-Williams, *The Shamans of Prehistory: Trance and Magic in the Painted Caves* (New York: Harry N. Abrams, 1998); David Lewis-Williams, *The Mind in the Cave: Consciousness and the Origins of Art* (London: Thames and Hudson, 2002). Anyhow, it tells a lot of the discussions that in 1930 Georges Bataille could still think himself as being the first to propose that the interpretations of primitive art should be made in connection with the differences in the picturing of animals and men in Paleolithic imagery (Georges Bataille, *Œuvres complètes*, tome I [Paris: Gallimard, 1971–1988], 254).
33. On Heidegger's position, see Part 1.
34. Barnett Newman, "The First Man Was an Artist," in *Art in Theory, 1900–2000: An Anthology of Changing Ideas*, ed. Charles Harrison & Paul Wood (Oxford: Blackwell Publishing, 2003), 576.
35. Ibid.
36. Ibid.
37. Rainer Maria Rilke, *Auguste Rodin*, trans. Victoria Charles (New York: Parkstone International, 2012), 15.
38. In a letter to Alfred Schaer dated February 26, 1924, Rilke explains, "That I was Rodin's secretary is not much more than an obstinate legend that grew up out of the circumstance of my once, temporarily, for five months, assisting him in his correspondence." Rainer Maria Rilke, *Letters of Rainer Maria Rilke*, trans. Jane Bannard Greene & M.D. Herter Norton (New York: W.W. Norton & Company, 1947), 333, accessed May 8, 2015, http://archive.org/stream/lettersofrainerm030825mbp/lettersofrainerm030825mbp_djvu.txt.
39. Rainer Maria Rilke, *The Notebooks of Malte Laurids Brigge*, trans. William Needham, 24, accessed November 19, 2016, https://archive.org/details/TheNotebooksOfMalteLaurids Brigge.
40. Rainer Maria Rilke, "In Celebration of Me (1909)", trans. Annemarie S. Kidder, accessed November 19, 2016, https://en.wikiquote.org/wiki/Rainer_Maria_Rilke. Translation modified.
41. Rainer Maria Rilke, *Duino Elegies*, trans. Stephen Cohn (Manchester: Carcanet, 1989), 65.
42. Ibid. Translation modified.
43. Ibid.
44. Ibid., 67.
45. Rilke, *Auguste Rodin*, 76.
46. Ibid., 22.
47. Ibid., 158.
48. Rainer Maria Rilke, "From *Letters on Cézanne*," in *Art in Theory, 1900–2000: An Anthology of Changing Ideas*, ed. Charles Harrison & Paul Wood, trans. Joel Agee (Oxford: Blackwell Publishing, 2003), 39.
49. See e.g. Marcella Lista, "Between Heaven and Earth: The Abstract Utopia of Franz Marc and the Artistic Theory of His Time," in *Chicago Art Journal* v. 10 (2000): 1–24.
50. Marc, cited in Luke Fischer, "Animalising Art: Rainer Maria Rilke and Franz Marc," in *Australasian Journal of Ecocriticism and Cultural Ecology*, vol. 3 (2013/2014): 52.
51. Ibid.
52. Ibid.

53. Ibid. "I seek to increase my sensibility for the organic rhythms of all things, a pantheistic feeling into the quivering and flowing of the blood of nature in the trees, in the animals, in the air—; to make this into a picture . . . I see no more fortunate means for the *animalisation* of art [*Animalisierung* der Kunst]' than the animal-painting ['*Tierbild*']."
54. Marc, cited in Lista, "Between Heaven and Earth," 3. It is well-known that Marc's interest in animal art had been influenced by Jean Bloé Niestlé. See e.g. ibid.
55. Rilke, cited in Fischer, "Animalising Art," 45.
56. Ibid.
57. Franz Marc, "How Does a Horse See the World," in *Theories of Modern Art: A Source Book by Artists and Critics*, ed. Herschel B. Chipp, Peter Selz & Joshua C. Taylor (Berkeley: University of California Press, 1968), 178–179.
58. Rilke, cited in Fischer, "Animalising Art," 47.
59. Theodor Däubler, cited in Peter Gay, *Weimar Culture: The Outsider as Insider* (New York: Harper & Row, 2001), 107.
60. August Macke, "Masks," in *Art in Theory: 1900–2000: An Anthology of Changing Ideas*, ed. Charles Harrison & Paul Wood (Oxford: Blackwell Publishing, 2003), 96; translation taken from K. Lankheit, ed. *Der Blaue Reiter Almanac,* English Version (London, 1974), 85.
61. Odilon Redon, "Suggestive Art," in *Art in Theory, 1815–1900: An Anthology of Changing Ideas*, ed. Charles Harrison, Paul Wood & Jason Gaiger, trans. Akane Kawakami (Oxford: Blackwell Publishing, 1998), 1066.
62. Sixten Ringbom, *Pinta ja syvyys: esseitä*, trans. Rakel Kallio, Esko Kukkasniemi & Paula Nieminen (Helsinki: Kustannusosakeyhtiö Taide, 1989), 52–53 & 100–101. As Ringbom points out, Rudolf Steiner and the theosophic paradigm played an important role in the rehabilitation of Goethean thought.
63. See e.g. ibid., 100; René Wellek, *A History of Modern Criticism: The Later Eighteenth Century* (Cambridge: CUP Archive, 1981), 208.
64. On the topic, see e.g. Ringbom, *Pinta ja syvyys*, 96–112. See also Wassily Kandinsky, *Point and Line to Plane: Contribution to the Analysis of the Pictorial Elements*, trans. Howard Dearstyne & Hilla Rebay (New York: Solomon R. Guggenheim Foundation, 1947), accessed November 20, 2016, https://archive.org/details/pointlinetoplane00kand and Paul Klee, "Ways of Studying Nature," in *Paul Klee: Dialogue with Nature*, ed. Ernst-Gerhard Güse (Munich: Prestel, 1991), 8.
65. Ibid., 100–101.
66. Ibid., 104.
67. Wassily Kandinsky, *Concerning the Spiritual in Art*, trans. Michael T.H. Sadler, 10, accessed March 21, 2015, www.semantikon.com/art/kandinskyspiritualinart.pdf.
68. Ibid. Translation modified.
69. Wassily Kandinsky, "The Cologne Lecture," in *Art in Theory: 1900–2000: An Anthology of Changing Ideas*, ed. Charles Harrison & Paul Wood (Oxford: Blackwell Publishing, 2003), 90; translation taken from Kenneth C. Lindsay & Peter Vergo, *Kandinsky: Complete Writings on Art* (Boston, MA: Da Capo Press, 1994), 394.
70. See e.g. the metaphorical usage of animals, and the concept of the animal at large, in F.T. Marinetti, "The Founding and Manifesto of Futurism," in *Futurism: An Anthology*, ed. Lawrence Rainey, Christine Poggi & Laura Wittman (New Haven, CT: Yale University Press, 2009), 49–53.
71. Kasimir Malevich, "From Cubism and Futurism to Suprematism: The New Realism in Painting," in *Art in Theory, 1900–2000: An Anthology of Changing Ideas*, ed. Charles Harrison & Paul Wood (Oxford: Blackwell Publishing 2003), 180; translation taken from T. Anderson, ed., *K.S. Malevich: Essays on Art 1915–1933*, vol. 1, trans. Xenia Glowacki-Prus, Arnold McMillin (Copenhagen) 19–41.
72. Malevich, "From Cubism and Futurism to Suprematism," 182–183.
73. Kasimir Malevich, "The Question of Imitative Art," in *Art in Theory, 1900–2000: An Anthology of Changing Ideas*, ed. Charles Harrison & Paul Wood (Oxford: Blackwell Publishing, 2003), 294; translation taken from T. Anderson, ed., *K.S. Malevich: Essays on Art 1915–1933*, vol. 1, trans Xenia Glowacki-Prus, Arnold McMillin (Copenhagen), 167–180.

74. Nevertheless, Kandinsky was not so hostile towards artistic observation of nature after all, nor was Klee. On Kandinsky and Klee in this sense, see e.g. Ringbom, *Pinta ja syvyys*, 108–109. Kandinsky was more and more assured in the 1930s that even abstract art and natural forms had to have some essential relationship: "Who knows," he rhetorically asked, "if all our abstract forms, every one of them, are 'forms of nature'?" Kandinsky, cited in Ringbom, *Pinta ja syvyys*, 109.
75. Piet Mondrian, "Plastic Art and Pure Plastic Art," in *Art in Theory, 1900–2000: An Anthology of Changing Ideas* (Oxford: Blackwell Publishing 2003), 390–391; translation taken from J. L. Martin, Ben Nicholson & Naum Gabo, ed. *Circle: International Survey of Constructive Art* (London: Faber & Faber, 1938).
76. See Jean Clair, ed., *The 1930s: The Making of the New Man* (Ottawa: National Gallery of Canada, 2008).
77. See e.g. Gill Perry, Francis Frascina & Charles Harrison, *Primitivism, Cubism, Abstraction: The Early Twentieth Century* (New Haven, CT: Yale University Press, 1993).
78. See e.g. Ringbom, *Pinta ja syvyys, 97–99*. Ringbom criticizes the attitude, while concentrating on C. H. Waddington, *Behind Appearance: A Study of the Relations between Painting and the Natural Sciences in this Century* (Cambridge, MA: MIT Press, 1970).
79. Alfred H. Barr Jr., *Cubism and Abstract Art* (New York: Museum of Modern Art, 1936), 12.
80. Mark Rothko, "The Romantics Were Prompted," in *Writings on Art*, ed. Miguel López-Remiro (New Haven, CT: Yale University Press, 2006), 59.
81. Marc, cited in Lista, "Between Heaven and Earth," 10.
82. George Maciunas, "Neo-Dada in Music, Theater, Poetry, Art," in *Art in Theory, 1900–2000: An Anthology of Changing Ideas*, ed. Charles Harrison & Paul Wood (Oxford: Blackwell Publishing, 2003), 729.
83. Rilke writes in 1922, in his letter to Margot Sizzo-Noris-Crouy, that "The poet must grow with the obligation that his *own* words should be separated fundamentally from the language of everyday life. In poetry, none of the words (and I'm speaking also of the tiniest of the words: from "and" to articles) is not *identical* with the words that are used daily: a regularity more pure, the decisive nature of the relations, the composition . . . all this makes the word to totally change its nature, makes it 'unutilizable,' worthless for everyday discussions." (Rainer Maria Rilke, *Hiljainen taiteen sisin: kirjeitä vuosilta 1900–1926* [Helsinki: Ntamo, 2012] 124–125.)
84. Michel Tapié, "From *An Other Art*," in *Art in Theory: 1900–2000: An Anthology of Changing Ideas*, ed. Charles Harrison & Paul Wood (Oxford: Blackwell Publishing, 2003), 629.
85. André Breton, *Manifestoes of Surrealism*, trans. Richard Seaver & Helen R. Lane (Ann Arbor: The University of Michigan Press, 1969), 18. "We really live by our fantasies when we *give free rein to them*."
86. Paul Éluard, *Œuvres complètes*, tôme 1 (Paris: Gallimard,1993), 44. My translation.
87. On the Surrealist conception of "psychic automatism," see Breton, *Manifestoes of Surrealism*, 26 & passim.
88. On the "olfactory hypothesis," see Nicholas Ray, "Interrogating the Human/Animal Relation in Freud's *Civilization and its Discontents*," in *Humanimalia,* vol. 6, no. 1 (2014): 22–28.
89. See Ray, "Interrogating the human/animal relation in Freud's *Civilization and its Discontents*," 21–28 & passim.
90. Sigmund Freud, "Civilization and Its Discontents," in *The Standard Edition of the Complete Psychological Works of Sigmund Freud*, vol. 21, ed. James Strachey (London: Hogarth Press, 1956–1974), 89. It must be added that Freud did not consider animal impulses revolutionary, as the Surrealists often did. The problem of animal societies and eusocial animals made Freud write: "Why do our relatives, the animals, not exhibit any such cultural struggle? We do not know. Very probably some of them—the bees, the ants, the termites—strove for thousands of years before they arrived at the State institutions, the distribution of functions and the restrictions on the individual, for which we admire them today. It is a mark of our present condition that we know from our own feelings that we should not think ourselves happy in any of these animal States or in any of the roles assigned in them to the individual" (ibid., 123). Nicholas Ray comments on Freud's remark that it may also be a

moment when Freud, with a characteristic gesture, may have started to doubt his own theory about the human-animal divide by imagining an animal condition in which the civilized life was "exemplified . . . in its absolute form: the termitary as the paradigm of civilization without discontents" (Nicholas Ray, "Interrogating the Human/Animal Relation in Freud's *Civilization and its Discontents*," 2014, 15).

91. Ibid., 31.
92. See ibid., 67.
93. Freud, "Civilization and Its Discontents," 96.
94. A theoretical text on the topic by a Surrealist writer, e.g. Georges Bataille, "L'érotisme ou la mise en question de l'être," in *Œuvres complètes*, tome XII (Paris: Gallimard, 1971–1988), 395–397.
95. See Freud "Beyond the Pleasure Principle," in *The Standard Edition of the Complete Psychological Works of Sigmund Freud*, vol. 18, ed. James Strachey (London: Hogarth Press, 1956–1974). For a thorough interpretation of Surrealist aesthetics based on Freud's ideas of Eros and Thanatos, see e.g. Hal Foster, *Compulsive Beauty* (Cambridge, MA: MIT Press, 1993).
96. Much could be said about the overturning of the ideal of Freudian psychoanalysis and its theory (as a theoretical literature attaching itself to the tasks of psychoanalytic therapy) in Surrealist philosophy (as an artistic theory hoping to give free rein to all of one's desires), the particularly Surrealist interpretation that attaches itself to the darker side of Freudianism, and grounds its thinking on the interrelations of Eros and Thanatos, see e.g. ibid.
97. Of Buñuel's admiration of Fabre in his own words, see Luis Buñuel, *My Last Breath*, trans. Abigail Israel (Glasgow: Fontana Paperbacks, 1985), 217.
98. J.H. Fabre, *Social Life in the Insect World*, trans. Bernard Miall, accessed February 12, 2013, www.gutenberg.org/files/18350/18350-h/18350-h.htm.
99. See Salvador Dalí, "L'âne pourri," in *"Il y aura une fois": Une anthologie du Surréalisme*, ed. Jacqueline Chénieux-Gendron (Paris: Gallimard, 2003), 183–186.
100. See Jacques Lacan, *Écrits: A Selection*, trans. Alan Sheridan (London: Tavistock, 1985), 1–7. On the parallelism between natural action and painting in Lacan's thinking, see Jacques Lacan, *The Four Fundamental Concepts of Psycho-Analysis*, trans. Alan Sheridan (London: Karnac, 1977), 114. "If a bird were to paint would it not be by letting fall its feathers, a snake by casting off its scales, a tree by letting fall its leaves? What it amounts to is the first act in laying down the gaze."
101. Roger Caillois, *The Edge of Surrealism: A Roger Caillois Reader*, ed. Claudine Frank, trans. Camille Naish (London: Duke University Press, 2003), 66–103.
102. Ibid., 66–81.
103. Ibid., 73n.
104. Ibid., 81.
105. Max Ernst, "An Informal Life of M.E. (as told by himself to a young friend)," in *Max Ernst* (New York: The Museum of Modern Art, 1961), 8.
106. Margot Norris, *Beasts of the Modern Imagination* (Baltimore, MD: Johns Hopkins University Press, 1985), 159.
107. Gavin Parkinson, "Emotional Fusion with the Animal Kingdom: Notes Towards a Natural History of Surrealism," in *The Art of Evolution: Darwin, Darwinisms, and Visual Culture*, ed. Barbara Larson & Fae Brauer (Hanover, NH: Dartmouth College Press, 2009), 267.
108. On the Surrealist interest in myths at large, see Didier Ottinger, *Surréalisme et mythologie moderne: les voies du labyrinthe d'Ariane à Fantômas* (Paris: Gallimard, 2002).
109. André Breton, "Fiches de préparation à l'Exposition," accessed November 19, 2016, www.andrebreton.fr/work/56600100071620.
110. See Caillois, *The Edge of Surrealism*, 110–123.
111. Ibid., 110–123. "Comparative biology should supply very valuable correlations as well, given that representation in certain cases replaces instinct, and that the actual behavior of an animal species can illuminate the psychological virtualities of man." (Ibid., 122.)
112. Ibid., 115n8; originally, in Victor Henry, *La magie dans l'Inde antique* (Paris: Ernest Leroux 1904), 242n1.
113. See Georges Bataille, *Œuvres complètes*, tome VII (Paris: Gallimard), 1971–1988, 598.

114. Ibid., 293–294.
115. Ibid., 291.
116. Ibid., 291.
117. Ibid., 292 & 295.
118. Ibid., 294.
119. The Surrealists and Bataille favored the free impulses of consciousness and poetry, while the Sartrean Existentialists had a cultural politics based on "prosaic literature" and "commitment." This was an important aspect of Bataille's theory in the 1940s, when it tried to oppose the supposedly utilitarian premises of the Sartrean theory. See e.g. Roni Grén, *Georges Bataille, historia ja taide* (Turku: Turun yliopisto, 2012), 90–107.
120. Bataille, *Œuvres complètes*, tome VII, 297–306.
121. Ibid., 293–294.

5 Conclusion

The Modern Other

The modern animal is above all *the other*. But to speak of the other is not an unambiguous task. This was proved by the theories of the avant-garde that thought that the other is just that which is to be affirmed by art.

My aim has been to show how the concept of the animal developed as such and what difficulties the concept encountered. I would next like to bring to mind the most important of the aspects of the development traced in the course of my work. The main arguments in progress in Parts 1 through 3 can be depicted through six different frames. After introducing them, I will return to the problems raised by the Expressionist and Surrealist solutions to the problem, and then consider what these solutions and this tradition still had to offer after World War Two.

Animalization of Art

1) *The thesis of human exception.* The arrival of philosophical aesthetics in the 18th century and the speculative[1] discussion on the nature of the arts that it supported prolonged and deepened the hegemony of the thesis of human exception. The tradition had to lean on the basic structures of the thesis by centering the discussion on the examination of the abilities thought to be exclusively human—among them, the ability to make art. It understood the nature of the abilities and the human-animal divide through *gnoseocentrist* and *dualist* proposals that affirmed the rupture between human and animal life.[2] In general, the speculative theories on the human-animal boundary and artistic aesthetics were also marked by the sort of *antinaturalism* Jean-Marie Schaeffer has traced[3]—*antinaturalism* given regardless of the value simultaneously proposed for the concept of the natural.

2) *Negative animal attribution.* Despite the fact that the major tradition, marked by the German philosophical aesthetics (the Kantian-Romanticist tradition) of the late 18th and the 19th centuries, has been favorable to the thesis of human exception, the interpretation of the human-animal divide it posited had its limits.[4] Its problem was highlighted by the questions through which it tried to understand the difference. The most important arguments in favor of the thesis were those supporting the idea of the unique origin and nature of human language and the related propositions about man's unique ability to be self-conscious, both of which supported the Romanticist idea that art was done in and by human measures. At large, the tradition has kept its appeal to many philosophical minds up to the present—mediated for example by the

Heideggerian tradition. As such, the arguments keep maintaining the basic structure of the thesis (the four principles defined by Schaeffer), but they confront problems in relation to the definitions of the animal when they consider human beings as essentially different. If the theory of the human exception was thus not airtight, it was mostly because of the negative nature of the determinations of the animal other: the various attributes given to animal life over the centuries started to have more and more connections with the formalist and revolutionary theories of art that identified the practice itself with dynamics of negativity.

3) *The visual world of animality*. In 18th-century France there opened a minor tradition that took up the possibility of thinking of the concept of animality in connection with imagicity, visuality, and even visual arts at large. I have tried to show the formation of the connection in mid-18th century context through the ideas held by Condillac, Rousseau, and Diderot. It should be noted that the 18th-century arguments concerning the connection were quite unstable, and that the human-animal boundary—despite some of the extremist proposals made by Diderot—was based on gnoseocentrism. Actually, the gnoseocentrist basis of the arguments was even the reason for such a connection. The connection was largely based on a collision of different—more or less politically motivated—discourses: the theoretical separation of, to use anachronistic terms, discursivity and imagicity in the politics of visual art, and the new ideas about the animal faculties of consciousness suggested by Condillacian empiricism, made the parallel between the animal world and the experience of visual art possible.[5] What was perhaps the most important property of the parallel between the two—this is my most important argument in connection with the topic—was the possibility of deducing formalist statements from it. As Michael Fried and Norman Bryson have showed, the statements resonated in the way authors spoke about the visual arts,[6] but the resonance has not seemed to have a large impact for the ways animals were pictured at the time.[7] Anyhow, it led to a situation that *could* be formulated as a dualism between the groups of concepts: (a) language–subjectivity–self-consciousness–human, (b) image–objectivity–the other–animal. These groups of concepts did not actualize their full potentiality in every way and in every discourse, but I do suggest that the connection became hugely relevant anew at the end of the 19th century, and attained its full bloom among the avant-garde that appropriated Nietzsche's work as part of their politics, in which they maintained the connection between the concept of the animal and the practice of art. Regardless of the differences between the two periods, the parallel between animality and visuality stayed at the core of an art theory, still affirmed from a similarly gnoseocentrist frame of thought in which the nature of visuality was understood through the opposition against which it was placed, that is to say, through its resistance towards human language.

4) *The ancestralized animal*. As Margot Norris has already suggested with her ideas on the biocentric tradition, the thesis of human exception was for the first time truly contested with the arrival of Darwinism.[8] Although Darwin was not a theorist of art, his impact was great in this area. I have tried to indicate the nature of his role by presenting some of the aspects of the theoretical adaptations made about his assertions on the human-animal divide. It should be noted that in this sense the rise of Darwinism is part of a much larger trend also in the history of art, a trend I have quite vaguely named as the objectivist paradigm. The passage cited from Darwin in Part 3 shows that Darwin himself might have had ideas about the nature of the arts and the experience of the beautiful that were ready to go far beyond the premises of the thesis

of human exception. However, it soon became clear that the ones adapting Darwin's visions to their theories on the nature of the arts were primarily interested in uniting both the gnoseocentrist basis and the Darwinist idea of natural selection. Theoretically, compromise between the two even seemed to be possible, but it soon became a rule among many of the "Darwinist attempts" that the difference between men and animals was affirmed almost only for the sake of declaring that the human experience of art could indeed deliver men to a state of superior knowledge—knowledge of their ancestral animal origins—an ability evidently denied the animal. In a paradoxical way, the post-Darwinist context meant that the animal now somehow stood both for and against man's self-consciousness. Anyhow, the animal had acquired a value related to man's, and his art's, own origin.

5) *The fictionalized animal world.* The crisis of allegorical symbolism[9] is often seen as being at the origin of artistic modernity. The process itself was long and far-reaching, and it makes no sense to sketch it out here with all the counter-reactions it initiated. At least from the early 19th century, its implications were felt.[10] They also had a huge impact on how, and on what grounds, animals themselves were pictured in the arts.[11] First of all, as the animal was stripped of most of its conventional symbolic qualities, it acquired them anew in a frame that more easily fitted the ideology of the avant-garde. Animals were not gods anymore, and hardly even images of the old pagan world. In spite of the metaphysical attractions of the old kind of symbolisms that some of the movements of the avant-garde embraced, there usually seemed to be a lack of symbolic value. The snake could have been the same one that deceived Eve into eating fruit from the tree of the knowledge of good and evil; the Sphinx still asked her questions of Oedipus—but their value was different for the simple fact that the metaphysical foundations of art and the concept of the animal were different too. Whether the context was naturalist or Symbolist, the purpose had changed. There was no going back to the pre-Revolutionary era.

It is evident that the role of animal painting itself changed drastically at the time. The 19th century did not only witness the sentimentalization of the genre of animal painting, a quality largely despised by the modern avant-garde, but also saw the birth of outdoor painting and the vogue of naturalist art. Although there was also an increase in texts and trends concerning animal politics and animal rights, these questions did not take up much space in the theoretical literature concerning the arts.[12] Rather, when the authors wanted to foster the depictions of animals in high art, they invoked different motivations. What is interesting, and also tells a lot about the functioning of the thesis, is that the depictions of the animals as themselves—that is, evoking them without an emphasis on their symbolic role—could perhaps better be defended on grounds that were subjectivist and fictionalized in their nature. I do not mean to say that the objectivist, realist, and naturalist tendencies would not contribute to the crisis—no one can deny their role as mediators towards the formalist premises of high modernism—but I think that the examples we cited from Delacroix's thoughts on art can give an important insight into the stakes at hand in the conceptual play around the images of animals (Delacroix evidently was not alone, and there are a multitude of writers defending the same position, but I have chosen his arguments since they can be backed up by his work).[13] The otherness of the animal (the concept of the animal) could now be represented, in the context of the deepening crisis of symbolism, by dramatizing and/or fictionalizing its world by means of an exoticism and supporting the hypertrophy of the beastly qualities towards immeasurability in human terms—often, in depictions of pain and violence. Actually, the representation of the value of the animal world—so important to Géricault's and Delacroix's lineage, the so-called French Romanticism—depended on these qualities, when it was offended on the one side by scientific objectivism, naturalism, and sentimentalism; and on the other by the need for human measures and more or less conservative symbolist tendencies.[14] The otherness represented by the estranged world of animality could as such be accommodated to the self-sufficient world of modern art.

6) *Animal Culture.* Friedrich Nietzsche's position is at the basis of my interpretation of high modernism, of the modernist avant-garde and its relation to the concept of the animal.[15] As Vanessa Lemm has pointed out in her book on Nietzsche's "animal philosophy," the difference between animals and men is for Nietzsche merely a cultural one.[16] But that does not mean that Nietzsche would leave the thesis of human exception completely behind. Nietzsche, in contrast with Darwinist naturalism, spares the dividing line if only to make it sound ridiculous.

Nietzsche's initiative is nonetheless very important, since he is able to meld many of the currents established during the preceding century into one coherent whole.[17] He also plausibly integrates the idea of the visual register in opposition to the conceptual, discursive one: "everything which distinguishes human beings from animals," he writes, "depends on this ability . . . to dissolve a picture into a concept."[18] And it is this connection that ties Nietzsche to some of the avant-gardist theories of his time. Nietzsche's perspective is indeed determinative in one sense, since he shows how the myth of the otherness of the animal can still be maintained, but also how the animal experience may at the same time be the ultimate model of artistic practice. This is done through an antinomy that Nietzsche posits between *culture* and *civilization*.[19] In an artistic world, milling around in the modern crisis, this is a liberating message: artists can testify to man's true animal nature by clinging to the formal (animal) qualities of their work. The qualities in question are indeed the ones that resist being put into the measures of man's conceptual thought—that is, for Nietzsche, qualities that can be discussed being visual in their nature. This sort of discussion is only possible in a modern world, where the dream of the purely visual has already entered a twilight. But for a moment, the artistic avant-garde and animality became, at least for some of the theorists following Nietzsche, political compatriots.

If Nietzsche's affirmational politics do not indeed step out of the frame determined by the thesis of human exception (it is still dualist and gnoseocentric, and its valuations are evidently antinaturalistic),[20] it does change the course of the avant-gardist view of the concept of the animal. Nietzsche's (and his followers') insistence on the merely cultural difference between man and animal, and on the importance of art in perfecting the return to one's true animal nature (basically, inverted human-animal dualism), even if profoundly absurdist in some of their consequences, called up a new kind of discussion on the topic. In only a few decades, its maintenance proved to be quite difficult.

The Formalist World of Creation

Let's say that the difficulties were not Nietzsche's. The problems only arrived when the ideas of the animal were transposed into the context of modern art and its practices; this is what we can learn of the parallel between Rainer Maria Rilke and Franz Marc.

As such, even the most enthusiastic of those who wished to animalize art had to notice quite soon the dead end that their views had constructed. It appeared as a total demolition of the symbolic quality, a demolition to which the followers of the line of thought had indeed attached the role of the image of the animal. The connection between animality and visuality, opened in 18th-century France, appeared empty in many areas.

1) *The empty word.* The first of the difficulties can be said to mark the whole tradition; at least, it is part of the thesis that connects animality with the visual register. The connection is structural, and it follows from the assumed otherness to which the animal is assigned. The problem follows from the positing of the *animot* to its role from which it represents the purely other of what is thought to be human—it makes the concept of the animal wholly arbitrary by its content and dependent on statements that try to define what is human in us. The animal becomes a floating signifier.[21] Animality can thus be understood as composed of almost any qualities whatsoever, as long as they do not threaten the place of man in the gnoseocentrist scheme. This is, one thinks, mostly made possible by the most extreme antinaturalist philosophies. But even when animality was marking a stage in our own evolutionary process, and even when art itself was somehow theoretically related to this animal origin (as in Haeckelian propositions) or to the need to depict animals (as in the example given to us by the interpretations of the cave paintings), one could just as easily forget the animal itself and use its image as an evidence of the superior means of the human mind. But as we saw with relation to the adaptations made by the pre-World War modernists progressing towards abstraction via animal images, it soon confronted a crisis when the value of the concept of the animal was perceived in its ambiguity and arbitrariness, hindered only by the formal restrictions of the given context.

2) *The empty eye.* Animal subjectivity also presented a problem. The desire to animalize art forced some of the artists and theorists to establish a dialectics between the inner and the outer also with regard to animal life. The idea was to form a theory to make animal life equal with the practices of visual art. This dialectics, especially important in Expressionist ideas, had then to be modelled according to the current beliefs of the nature of the gaze of the animal—in principle, the animal gaze *in general.* This followed the guidelines of a long tradition that had learned to think of the animal gaze opening onto the world without the restrictions and advances posited by human language and conceptual thinking. As man's relation to the world was thus posited as that of a perceiving subject, one that could comprehend what it saw in terms of the basic categories of human understanding, the visual artist could conversely imagine the animal as a subject that sees nothing in a human sense,[22] a subject whose gaze was purified of the load of history. This, through circular reasoning, made the animal once again parallel with the needs of the avant-garde. It gave an answer that the artist was already seeking: if the animal condition was being sought, the concrete forms should be made imperceptible to the human gaze. For this reason, one needed to suspend man's objectified relation to the world. But this, in its turn, produced a paradox intriguing the mind of the artist whose work was nonetheless to make objects of perception—perhaps solvable only through force, and in the theoretical world determined by dialectics.

3) *The empty space.* The development did not have to lead suddenly into an abstract work of art. Marc himself hesitated. But there was no hindrance to the development

either. In other words, the concept of the animal that the pre-War Expressionist discourse supported did not give any reasons for not demolishing the representative function of art in its totality if one wanted to animalize art. This development perhaps uncovers something essential about the peculiar progression of Expressionism towards the forms of abstract art. Perhaps Kandinsky himself had taken these steps, and only much more quickly than Marc, since his art had developed towards pure abstraction through the mediation of a short period of animal paintings. However, the animalization of art, as Marc wanted to call his project, was paradoxically perfected only by an abandonment of all animal motifs and all references to the animal sphere of life, to all objects with which man could understand animal life. One could as well, if one wanted, leave the image of the animal world totally behind and continue making abstract compositions, from then on motivated by the need to mystify human life. The human spirit had once again overcome its objectified status! Ideally, in the name of this reasoning, animality was perhaps at the same time achieved and abandoned at the moment of their art's perfection. Nietzsche's question about the animal that would have said something if it only would not have forgotten what it wanted to say[23] had perhaps unintentionally been reflected in the mirror made available by abstract art.

It almost seemed that the whole modern tradition of thinking the concept of the animal had been somehow concentrated on the question of abstract art during these years (I know that this sounds overstated, but I want to make clear how far the crisis related to the concept had gone): the need for a transcendent visual object, the conception of the animal as the pure other, the abandonment of the representative function, the apotheosis of the inner visual register, the revolutionary potential of the dream of the purified gaze and its connection with the concept of the animal. Once these had been brought together by artistic means, the dream of an animal culture was not the same anymore; it had been drained of its meaning. It is no wonder then that the spark that it had given to some of the artists faded away fairly soon.

The Surrealist Solutions

It is beginning with Surrealism that the symptoms of the modern attitude are brought on stage more explicitly, even in the sense in which contemporary culture has learned to speak of them. The movement took part in the discourse on the concept of the animal by integrating its idea of animality into its practices of art, but it also managed to introduce critical stances from inside the paradigm. I have attempted to present the history of this integration and criticism in the interwar period in three different stages, which of course overlap with each other and could always be only ideally present. However, they are determined by the topics on which the writers and artists laid emphasis in their use of the concept of the animal. The history of the movement makes an important case in point, since Surrealism was perhaps the only large avant-garde movement between the wars that took seriously the task of interpreting animality and using animal images systematically in the artworks it engendered.

1) *Dream.* In Surrealism, the dream is always political; the dream of the animal world was supposed to open man's consciousness to a world in which the bourgeois conception of work was not dominant. In the early years of Surrealism, the animal was

mainly taken as a representative of the world outside human struggle for recognition, an idea of a world that was formed through Nietzschean-Rilkean trails. The animal presented the pure natural drives. The Surrealist theory, however, was charged with sexual implications, its interpretation of symbols being also governed by Freudian ideas.[24] In this sense, animal images represented, and could be made to encourage, our need to break out of the sexual constraints imposed on us by Western civilization. This meant that "the empty sign" was now filled with meaning up to the point of overdetermination. The politicization of the image of the animal—constructed on a similar basis as the earlier critical statements against the culture of representation and against the fragmentary nature of modern life—was now straightforwardly posited against the bourgeois domination of human and animal drives. Thus confronted, the Freudian interpretation, not unexpectedly, changed its course towards darker terrains in short order.

2) *Violence.* As manifold as the Surrealist use of the image and the concept of the animal proved to be, one can trace a development from the early years of the 1920s towards the more somber world represented by the animal in the turn of the next decade. Evidently, the turn was catalyzed by the observation of the problems inherent in the Nietzschean-avant-gardist lineage. Max Ernst's *œuvre* is in this sense the most telling example.[25] Ernst's identification with his birds also tells its story about the quest towards animality under the conceptions supported by the Surrealists. This history can also be interpreted against the background of the popularity of anthropological questions at the time, and with the so-called rehumanization of culture.[26] The violence of the animal world was now, beginning from the late 1920s, being evoked in Surrealist politics again and again, because of its ability to mirror our own animal impulses constantly at work inside us, but simultaneously, it still carried with itself the dream of breaking from the constraining chains of current human civilization. What was new was the forceful emphasis on the terrifying powers lying below the futile surface of civilization: as such, the themes of fiction and violence were discharged through animal images because of the revolutionary purity they could impose against the alienated bourgeois world, though not without sacrifices being made. This abject world, that now arose from the borderline of humanity, was just what they had found in the early 1930s in the downfall of human-animal dualism in poetic thought.[27]

The change is evidently a result of the internal dynamics of the Surrealist movement and of the changes within the interpretation of the Freudian paradigm, but it must also be accepted that the Surrealist interest in the natural sciences of the time played a role in them—Jakob von Uexküll's biosemiotics can in some sense be called a more moderate, non-Surrealist counterpart of Caillois's early texts.[28] Moreover, the dualist frame through which animality had been understood until then was in a crisis as a result of these changes, and the views imposed on animality among the Surrealist group at this time can also be seen as marked by the need to find better answers without compromising the dualism that was still important in the revolutionary theories supported by the politics of Surrealism. Perhaps the biggest problem was then the inability to draw the line between animality and humanity clearly (which the use of the animal in this regard still required)—a condition that was nonetheless perceived as essential for the practice of defining the basest of man's impulses.

3) *Myth*. Anyhow, with the dividing line being questioned on many fronts, there was no turning back; once again, the line could be conserved only by fictionalizing it, and explicitly defining its value as positive. This is the frame in which Georges Bataille spoke of a "poetic lie of animality," which he conceived as the major support of all human culture, especially of religion. Without examining further Bataille's thought, one could see its counterparts in many Surrealist works of the late 1930s and after. The thesis also makes clear how constitutive the role of the animal, from time to time, had been present in Surrealist thinking on the arts, but it also brings to mind how the sort of interpretations centered on the concept of the animal, the *animot*, worked. Bataille's lucid remarks can also be seen as mirroring the interest he felt towards the origin of the arts in the cave paintings found from the Franco-Cantabrian region. As I showed in one of the subchapters of my work, the interpretations quite easily rearranged the reasons for making animal paintings in favor of the human-animal dualism, now represented by the myth of the animal world. In other words, the nature of the image of the animal—e.g. the species of animal depicted—did not matter when facing the fact that the image itself could be considered to represent animality *at large* and *in general*. The Surrealists, embodied here by the somewhat unorthodox texts of Bataille and Caillois, took a new step when they brought into their art and their theories the self-reflexive understanding of the structural and the mythical nature of their own dreams of animality. Animal art could now be made from a self-imposed perspective towards the nature of the myth—perhaps the modern myth *par excellence*.[29]

Indeed, the Surrealist interest in myths, in the late 1930s and 1940s can above all be seen as a result of the kind of disillusionment that the group was going through with its cultural politics and lost hopes for the projects to which the ideology of the movement had attached itself. But be that as it may, it also marks an important point in the history of the modern conception of the animal in the artistic field. In some sense, Bataille's theory tells the story of a lost innocence. For Nietzsche, animal life is still sensed as a possibility, and art is given as a medium with which to grasp man's own and true animality. For Bataille, the whole idea of animality is only given as a poetic fallacy, though as one of the most precious kind—as a cornerstone of human culture. Bataille's sublime definition of the fantasy does not only give us a proof of the important status of the concept of animality among at least some of the Surrealists, but it marks the moment when the discourse was able to reflect on its own conditions.

Leaving the vast polemics aside, the primitivism of the 20th century nonetheless coincided with the increasing interest in seeing art anthropologically, and this tradition was embraced by Bataille and the Surrealists, as well as many of their opponents. Indeed, the 1920s and 1930s had already witnessed, following Devin Fore, "a rehumanization of art" that returned man to the epistemic center and "enthroned anthropology as a master discipline among the cultural and natural sciences."[30] This development, turning towards questions asked anthropologically, had mixed effects: it carried within itself the seeds of a transformation, since the questions asked about the origin of art were once again already being asked in relation to animal nature. There were no unequivocal answers to these questions. And in any case, the answers differed greatly. The origin, in one way or the other, managed to maintain the label of a paradise lost. But the dualism, which established the origin as a paradise lost, while fastening itself anthropocentrically to that which it thought being human, had its problems: if art meant in the context of modern avant-garde that one had to step, within the human practice of art, beyond what was human, the demand established once again a connection to such concepts that were defined in advance as representing something outside of humanity. It is no wonder that Paul Celan called for a spiritual realm of the uncanny "turning towards human," in which "the monkey, the automatons and with them . . . oh, art, too, seems to be at home."[31]

The Animal Itself

To continue with generalizations, the valuations of the concept of the animal were mainly determined, from the 1930s on, by whether the movements of the avant-garde took a formalist or an anti-formalist stance. This is of course not the whole truth, but mostly this was a dominant factor. While the anti-artistic movements continued to define animality in positive terms, the more strictly formalist movements often had less to say about animality.[32] In this light, it is no surprise that the theorists of the postmodern animal have largely stigmatized the modern era for its incapability to consider "the animal itself"; its severity in the Anglo-American context may also be a consequence of the history of the mid-20th century American theory, in which the concept of the animal seems to have played an even lesser role than in Continental developments. However, the hegemony of Greenbergian modernism[33] and American abstract art seems to have totally dismissed the space reserved for the animal in some European discussions. On a theoretical level, this has ensured the animal's status in postmodern theory as a symbol of resistance, at least in the sense in which the history of modern art can be seen one-sidedly as the history of medium-specific practices. But it must be added that direct resistance against Greenbergian modernism—the conceptual artists and writers following more or less in the footsteps of Duchamp—did not contribute much to the discourse on animality either. The linguistic turn, and the art theories that it launched, evidently had little reason to talk about the animal in the proper sense.

The heritage of Surrealism, the uncanny realm of the art of the insane, the naïvist revival, or the revolutionary potential of the outsider culture was indeed perhaps the arena in which the animal could still pasture, even after the war. Evidently the most important of such movements were *art brut* and the short-lived CoBrA movement, which also moved towards the mediation of Surrealist pictorial politics into Situationist cultural practices. Asger Jorn was especially keen on portraying animals—animal species more or less imaginary—in his works. Jorn also presented a theoretical interest on the topic, acknowledging his inheritance of the tradition that criticized the symptoms of civilization through animal means. Nevertheless, unlike many of his contemporaries, Jorn was even able to show the way forward to discussions of the human animal and hope to arrive, after high modernism, "at symbols common to all," creditably by returning to the pictorial practices of the long tradition of popular culture and its ways of seeing the animal as an underdog. In 1950, Jorn explained his reasons for including manifold animal figures in his works in a letter he wrote to his fellow artist Constant (Nieuwenhuys):

> Popular art is always fantastic and symbolic. There is Grandville and Walt Disney and there is us. Often one can better describe the essence of struggles among men by means of primitive, simple, fantastic beasts, raw instincts, than by painting an

> individual situation, a battle between police and striking workers . . . It is necessary to surpass this individualism in art and arrive at symbols common to all . . . We must not describe the human animal, but rather describe ourselves as human animals.[34]

This is mostly the strategy that the political Surrealism of the early 1930s had adopted. It is true that after the war, the then-to-come Situationist politics at large—and Jorn among them—tried to overcome the obstacles that followed from Surrealism's ineptitude at commitment and flexibility, which they thought intermittently gave it a weakness that led to political impotence. However, these unstable political paths had already given the animal a critical value which the following movements could scoop up. The legacy of understanding man as a human animal, and of mirroring man's habits in relation to the animal through anti-bourgeois political stances, is the legacy that Surrealism left, and which has even given it a sort of pro-postmodern aura in contemporary theory.

An approach to animals that can be seen as a continuation of the reflective attitudes of some of the Surrealists was also embraced by some remarkable artists of the following decades. At least two of them have managed to make such important contributions to the tradition that, in my opinion, they can't easily be set aside while discussing the relation between contemporary theory and modernist ways of considering the animal: Joseph Beuys and Marcel Broodthaers. Both of the artists, children of their time as they were, reacted significantly against the demands of medium specificity—a tradition at least as old as the concept of *fine arts*—which had erected the prohibition against exhibiting animals in a space reserved for artwork. At the same time, they testify to the increasing interest among the artists to comment on the status of the animal and the natural life in their artworks at the borderlines of modernity.

Beuys had made animal art in different media already as the 1970s approached, and his influence on the tendency to integrate ecological questions in art gallery space is immeasurable.[35] In my opinion, the most important of his works, in the sense of representing the animal itself within the practices of modern art, is the performance called *I Like America and America Likes Me* (1974), during which Beuys spent three working days in a gallery space with a wild coyote.[36] Beuys was not the first to transgress the modernistic law forbidding the exhibition of real—whether dead or alive—animals as art; Iannis Kounellis had already made his installation *Untitled (12 Horses)* in 1969. But it is Beuys's work that rightly enjoys the higher status in the tradition of thinking of animals themselves as active subjects within the exhibition space, a tradition that has grown during the last decades. The two works also present the two sides of the current of exhibiting real animals in the context of art: in Kounellis's work, the horses are there to be gazed at in their otherness, while Beuys approaches a point-of-view according to which the human-animal relation—even with a wild animal—is supposed to be interpreted in a frame determined by mutuality. Perhaps the coyote in Beuys's performance is still an object—still an art object when exhibited, not living its so-called natural life[37]—but as I see it, Beuys's performance indeed stands at the threshold of any such questions that the modernist discourse on art and animals was ever able to ask.

Broodthaers, in his turn, has been acknowledged for decades as having been a key figure in making institutional criticism by the means of art, since he has cleverly been playing with modernism's art practices. However, while this side of Broodthaers's art has become well-known, it has oddly been overlooked—even regardless of the fact that

Broodthaers himself made this quite explicit—that animal imagery held a privileged place in this critique for years.

From Broodthaers's early works, from his collection of animal poems called *Pense-bête* (1963–64) to the series of installations for which he is famous—*Musée d'Art Moderne* (1968–1972)—animals indeed possessed a central role in his most characteristic works. What is more important, they took a more or less analytical attitude towards the meanings that animality had had in the art world. The name of the collection of poems, *Pense-bête*, may have already testified to the fact: it may allude to Lévi-Strauss's idiom that "animals are good to think with,"[38] but while the work has also been catalogued as being part of Broodthaers's visual artworks, with its covers permanently closed (cast in plaster),[39] it may as well be connected to the degradation of animal imagery (its segregation from the high arts); or even to connect itself with Marcel Duchamp's—an artist to whom Broodthaers explicitly referred in some of his works—explanation of his abandonment of the art of painting because he didn't want to be called "bête comme un peintre."[40] In any case, soon after exhibiting the plaster-cast book as a work of art, Broodthaers turned to making a series of works filled with real mussels, works which have been interpreted, in continuation with the artist's later installations, as commenting upon the Belgian identity of the artist, and of course, this identity is labeled through animal imagery.[41] Whatever the mussels had to say, the analysis of the role of animal imagery, and the partial covering of the role in modernist art practices, was a topic that kept Broodthaers very busy.

"The eagle is," wrote Broodthaers, commenting upon his *Département* of "modern art," "like a paper tiger, a sickly monster. It nests in public museums."[42] Without a need to enumerate all the steps that Broodthaers needed to go through to make such a work of

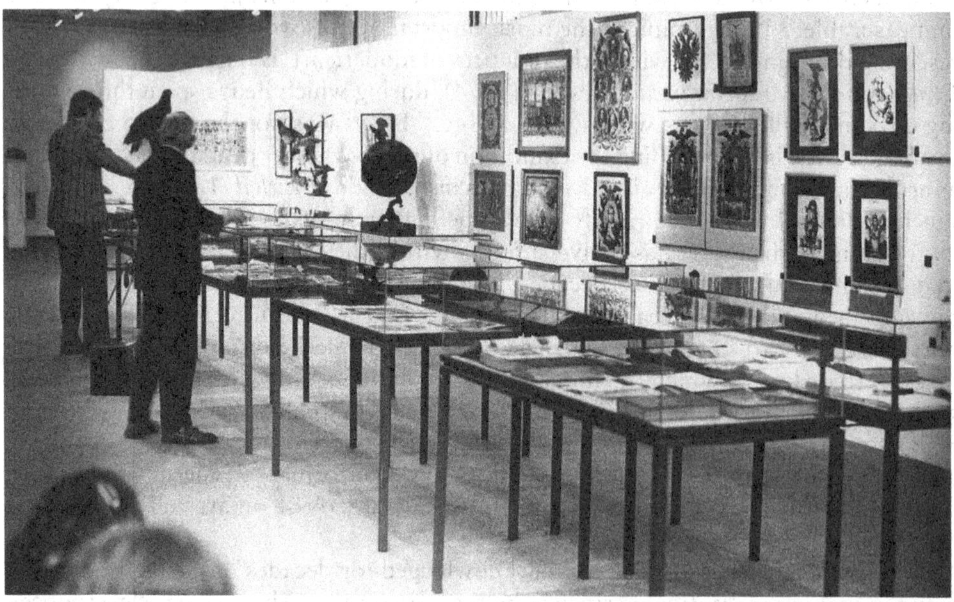

Figure 5.1 Marcel Broodthaers, *Musée d'Art Moderne, Département des Aigles, Section des Figures. Düsseldorf Kunsthalle, May 16–July 9, 1972.* © Kuvasto 2016

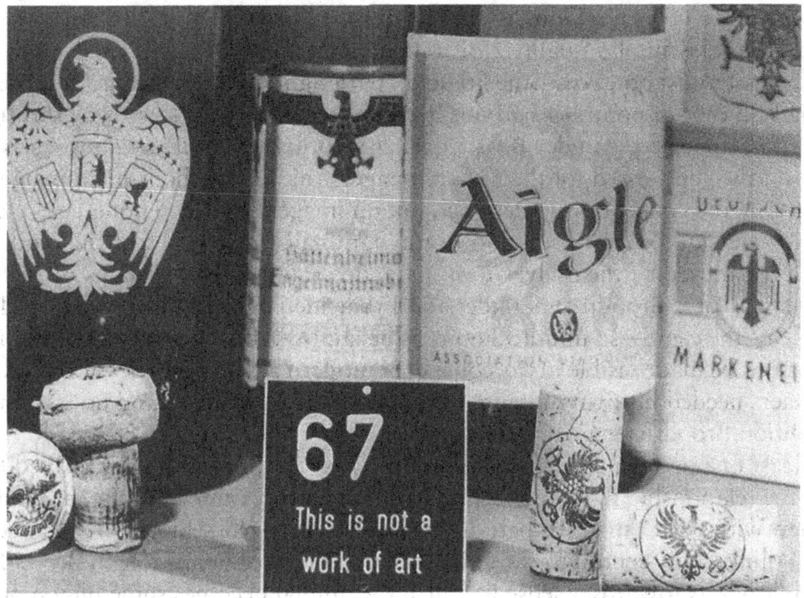

Figure 5.2 Marcel Broodthaers, *Musée d'Art Moderne, Département des Aigles, Section des Figures. Düsseldorf Kunsthalle, May 16–July 9, 1972.* © Kuvasto 2016

art (the artist's "own museum" as an artwork), beginning from the opening ceremony, at which there was no works to be seen, what interests us is the multitude of images of eagles, in the "Section des figures" of Broodthaers's museum, that the artist decided to show, with a careful notice placed in front of each of the images reminding the visitors that the images were "not works of art." The last of the gestures was, it seems, only a call for a critical reaction and the need to understand how animal imagery had molded our readings of the artworks. Broodthaers himself explained the role that the eagle played:

> It puts art into question through the art object that is eagle. Eagle and art are merged here. My system of inscription and the general atmosphere that results from the repetition of the object and the confrontation with projections of publicity invite viewing the art object, that is to say, an eagle, viewing an eagle, that is to say, an art object, with a truly analytic perspective—that is to say, separating in an object that which is art and that which is ideology. I want to show ideology such as it is and avoid art from making ideology invisible—that is to say, efficient.[43]

What is totally new in Broodthaers's installation is not the criticist stance towards the art institution, a criticism made to act from the inside (in Broodthaers's own country, the Surrealists had made such practices known long ago), but the way in which the imagery was put on stage, and the reasons for making such a gesture. The animal here starts to habit a terrain that it had not been able to reach within the boundaries of the modernist discourse on the arts because of its idealistically determined role as the human other, a role that consequently had meant that the animal was either thrown out of the sphere of art, or, in other contexts, had been taken as an *exemplum* of a life outside linguistic determinations.

Although the symbolic qualities of animal imagery had been raised many times within the modernist practices—the Freudian paradigm and its Surrealist interpretation might make a case in point—it had not been understood in the same way through the context of its appearance under Broodthaers's scrutiny. *Département des aigles* was a museum of the suppressed, of the animal imagery and its symbolism; an imagery not seen as fitting the concept of modern art, but an imagery that had nonetheless kept its fascination outside the museum, and in the nationalist discourses of the modern era to which the museums themselves took part, determining our readings of the works in their historicity. Broodthaers understood that ideological implications stuck onto the image of the eagle—so multitudinous as he showed them being—could not for the same reasons be made visible in the context of modern art—as "works of art"; for this, Broodthaers needed his own museum of "modern art," No more and no less, this was the condition Broodthaers managed to put on stage.

In at least one performance given in the *Département*, Broodthaers affirmed his critique of the relationship that the theorists had had with the animal. This was done in an "Interview with a Cat" that was part of the installation in Düsseldorf in 1970. The interview is, I think, quite central to the understanding of the animal's role in Broodthaers's museum: it points towards a sense in which the animal becomes something more than just an arbitrary part of Broodthaers's institutional commentary. In the work itself, the artist himself interviews a cat by demanding, among other questions set up to ridicule the art world, that the cat answers René Magritte's famous enigma, posited in Magritte's painting *The Treachery of Images* (1928–29), that the pipe represented in an artwork "is not a pipe." The cat does its part very well: Magritte's pseudo-philosophical critique and its foundations appear with their relations to the animal. An animal life, often supposed to be seen as a symbol of a world beyond representations, loses its evident value in the interview, when the cat effaces the critique and speaks for itself. It does not matter if the cat doesn't understand the critique. The main point is that the animal is shown through its reactions towards linguistic representations, producing them at the same time. Perhaps we start to understand the cat more, within the boundaries of the interview, than we do Broodthaers's insistence itself. A short excerpt will make the point clear:

Marcel Broodthaers:	Is that one a good painting? . . . Does it correspond to what you expect from that very recent transformation which goes from Conceptual Art to this new version of a kind of figuration, as one might say?
Cat:	Miaow.
MB:	Do you think so?
Cat:	Miiaaw . . mm . . miauw . . miauw.
MB:	And yet this colour is very clearly redolent of the painting that was being done in the period of abstract art, isn't it?
Cat:	Miaaw . . miaaw . . miiaw . . miaw.
MB:	Are you sure it's not a new form of academicism?
Cat:	Miauw.
MB:	Yes, but if it's a daring innovation it's still a contestable one.
Cat:	Miaw.

MB:	It's still . . .
Cat:	Miaw.
MB:	Er . . . It's still a matter of markets . . .
Cat:	Miaauw.
MB:	What will the people who bought the previous things do?
Cat:	Miauw.
MB:	Will they sell them?
Cat:	Miiauw . . mia.
MB:	Or will they continue? What do you think? . . . Because, at the moment, a lot of artists are wondering about that.
Cat:	Miaauw . . mm . . mii . . miauw . . maaw . . miaauw . . iaw . . mm . . Miauw . . miauw . . MiAUW!
MB:	In that case close the Museums!
Cat:	MIAUW!
MB:	This is a pipe.
Cat:	Miaouw.
MB:	This is not a pipe.
Cat:	Miaouw
MB:	This a pipe!
Cat:	miAOUW
MB:	This is not a pipe!
Cat:	miaouu
MB:	This is a pipe!
Cat:	miaOUUW
MB:	This is not a pipe!
Cat:	miaouuw
MB:	This is not a pipe?
Cat:	miaw[44]

Is Broodthaers here referring to E.T.A. Hoffmann's Murr the Cat, whose "Miaouwing" Hoffmann had called superior to human speech, since the cat was able to express so much with one word?[45] Or is he turning Michel de Montaigne's pensive question, "When I play with my cat, who knows if she is making more a pastime of me than I of her?"[46] into a question "of making art with my cat"? Or is he just trying to show, by repeating endlessly Magritte's pun, that one could imagine the human–animal interview continuing eternally without the interviewer and interviewed ever able to say much to each other on the subject? The latter seems to have been the conclusion to which much of the modernist paradigm could always return.

Notes

1. By "speculative theory of art," I mean the tradition of aesthetics beginning from the time of Kant. "Speculative aesthetics," in this form, is a term provided by the works of Jean-Marie Schaeffer. See Jean-Marie Schaeffer, *Art of the Modern Age: Philosophy of Art from Kant to Heidegger*, trans. Steven Rendall (Princeton, NJ: Princeton University Press, 2010).
2. See Part 1. The two terms are borrowed from Schaeffer's analysis of the Western philosophical tradition and its support for the thesis of human exception. Jean-Marie Schaeffer, *La fin de l'exception humaine* (Paris: Gallimard, 2007), 27–29.

3. Ibid., 29.
4. It should be noted that the role of the animal agency in Romanticist philosophies is not so easily swept aside. My argument, however, is that the animal was thrown aside from the considerations of art by giving it a role that was negatively determined—as a *non-human* agency.
5. The place of empiricism in the development should be noted, although I have mostly omitted it from my considerations, because the context of empiricist criticism is beyond the scope of this book. What is essential for my argument is that in France, the empiricist currents had turned the theorists of art away from the techniques of representation towards the experiences given by art. See e.g. Jacqueline Lichtenstein, *Les raisons de l'art. Essais sur les limites de l'esthétique* (Paris: Gallimard, 2014).
6. Michael Fried, *Absorption and Theatricality: Painting and Beholder in the Age of Diderot* (Chicago: Chicago University Press, 1980); Norman Bryson, *Word and Image: French Painting of the Ancien Régime* (Cambridge: Cambridge University Press, 1981). I thank Bryson's work for noticing the 18th-century needs and techniques to overcome the allegoric function; Fried, in his turn, has shown how far beyond figurative matters and towards modern formalist means the politics of the gaze, suggested by the texts of Diderot, could have led artists.
7. My analysis, in Part 2, touches upon a quite narrow scale of material. My idea has only been to show that the possibility of connection between animality and this sort of imagicity was perceived at the time. As far as I can see, it did not change the ways animals themselves were pictured, although I have indeed alluded to the particular case of Théodore Géricault, following the analysis made of his works by Fried. If such changes, connected to the conceptions of the animal, ever occurred in a large scale, they have not appeared to me in the course of my research. A comprehensive study of the changes of the paradigm in France at the turn of the 19th century regarding the image of the animal has yet to be done.
8. Margot Norris, *Beasts of the Modern Imagination* (Baltimore, MD: Johns Hopkins University Press, 1985).
9. The crisis was of course quite manifold, and it also progressed in many ways; it should not be understood only as a uniform progress towards non-representative art. One of its most notable functions was the separation of the concepts of the symbol and the allegory. In this accord, it should be noted that, as Frances S. Connelly seems to have observed in relation to the development of the modern conception of the primitive, before the symbol was released from most from its allegorical tasks, as in the late-19th-century trends of philosophical symbolism, allegorical and emblematic figures had first presented themselves in some Romanticist currents as alternatives to the classical hegemony of narrativization in (history) painting. Connelly analyzes Philipp Otto Runge's work in this regard in particular. See Frances S. Connelly, *The Sleep of Reason: Modern European Art and Aesthetics, 1725–1907* (University Park: The Pennsylvania State University Press, 1995), 44.
10. In my opinion, they had already started to be felt in the preceding century, but mostly in France. This much I have taken from Fried's and Bryson's works, although the two disagree with each other on the nature of the change that took place.
11. It should nevertheless be noted that at the time animals were also discussed in the writings of the theorists of the nature of the symbolism in the arts. See for example Richard Payne Knight, *An Inquiry into the Symbolical Language of Ancient Art and Mythology* (New York: J.W. Bouton, 1892).
12. In this sense, the ones who have criticized modern attitudes for the lack of interest towards animals themselves are right.
13. As we have seen, at least the Brothers Goncourt and Théophile Gautier have given us examples to cite in this regard. On Gautier, see Théophile Gautier, "From Preface to *Mademoiselle de Maupin,*" in *Art in Theory, 1815–1900: An Anthology of Changing Ideas*, ed. Charles Harrison, Paul Wood & Jason Gaiger (Oxford: Blackwell Publishing, 1998), 100.
14. It should be noted that in the counter-reactions against the crisis of symbolism, animal images were frequently used. These reactions were often tied to the premises of the thesis of human exception, by clinging to the idea that these symbols had a privileged place in producing and affirming man's self-conscious nature. We know that these sorts of ideologies also dominated the Symbolist currents at the end of the century. Retrospectively, the ancient

symbols—the images of animals with them—were *formally* considered carriers of the wisdom of the old world, when the modernist developments were criticized. As we saw in Part 3, Delacroix also severely criticized the naturalist tendencies of the time for their superficial means of resorting to formal means adopted from scientific objectivity.

15. I do not mean to say that all modernists were Nietzschean, but I wish to point out that his ideas of the connection between animality and visual arts were approved by important modernist tendencies that aimed to consolidate, for one reason or another, formalist politics of high modernism with the appropriation of the concept of the animal, and that regarded animality as a positive value.
16. Vanessa Lemm, *Nietzsche's Animal Philosophy: Culture, Politics, and the Animality of the Human Being* (New York: Fordham University Press, 2009).
17. Some traditions do indeed stress the asystemic nature of Nietzsche's thought, also with regard to his texts of the animal. In any case, I have wanted to make clear in my text, defending Lemm's work on the topic, that Nietzsche uses the concept of the animal quite coherently throughout his career.
18. Friedrich Nietzsche, *The Birth of Tragedy and Other Writings*, trans. Ronald Speirs (Cambridge: Cambridge University Press, 1999), 146.
19. The antinomy between *culture* and *civilization* is thought by Lemm to be the constitutive pair of concepts through which Nietzsche formulates his conception of the animal. See Lemm, *Nietzsche's Animal Philosophy*, 10–30.
20. The aim of culture is, according to Nietzsche's Romantic line of thought, "to educate oneself . . . against one's self, to a new habit and nature, out of an old first nature and habit" (Nietzsche, cited in ibid., 24).
21. I am using the term "floating signifier" as it is understood in Jacques Lacan's thinking, as a symptomatic and privileged object which can be made to adopt contrasting meanings.
22. Georges Bataille's theory of the myth of "the poetic lie of animality" explained the idea in these terms. According to Bataille, the idea of an animal world can—because of the fantasy of otherness that supports our conceptions of animality—only be evoked as a "vision, which gives us *nothing* to see" (Georges Bataille, *Œuvres complètes*, tome VII [Paris: Gallimard, 1971–1988], 293).
23. Friedrich Nietzsche, *Untimely Meditations*, trans. R. J. Hollingdale (Cambridge: Cambridge University Press, 1997), p. 60 §1.
24. The Surrealist interpretation of Freudianism has been under heated discussion ever since. In any case, it seems evident that it started to chart much darker terrains at the end of the 1920s. Although it is true that Freud too had started to turn in a different direction, the Surrealists were at odds from the beginning (although not always conscious of it) with the means and ends of Freudian psychoanalytic discourse; how else, when Surrealists thought that the powers of the unconscious had to be liberated completely to make a new Surrealist society, an idea was hardly conciliable with the ends of Freudian therapy. On this topic, see e.g. Hal Foster, *Compulsive Beauty* (Cambridge, MA: MIT Press, 1993).
25. Ernst's thoughtful use of animal images has been noticed by many. On Ernst's identification with birds, see Werner Spies, *Max Ernst—Loplop: The Artist in the Third Person* (New York: George Braziller, 1983). On the development of the Nietzschean discourse in Ernst's art, see e.g. Norris, *Beasts of the Modern Imagination*. Of the differences between the two, Norris writes, "Where Nietzsche 'forgets' the 'other,' cultivating the genuine oblivion of animals toward the consciousness of others as subjects, Ernst practices instead a cunning, and sometimes violent, attack on the 'other'." (Ibid., 135.)
26. See Devin Fore, *Realism after Modernism: The Rehumanization of Art and Literature* (Cambridge, MA: MIT Press, 2012).
27. Interestingly, this is just what Julia Kristeva seems to have found a few decades later when she declared that "on close inspection, all literature is probably a version of the apocalypse that seems to me rooted, no matter what its socio-historical conditions might be, on the fragile border (borderline cases) where identities (subject/object, etc.) do not exist or only barely so—double, fuzzy, heterogeneous, animal, metamorphosed, altered, abject." Julia Kristeva, *Powers of Horror: An Essay on Abjection*, trans. Leon S. Roudiez (New York: Columbia University Press, 1982), 207.

28. On Uexküll's theory, see e.g. Jakob von Uexküll, *A Foray into the Worlds of Animals and Humans: With a Theory of Meaning*, trans. Joseph D. O'Neill (Minneapolis: University of Minnesota Press, 2010).
29. An idea that many of the Surrealist artists put into practice. Most notably this happened in the 1947 exposition of Surrealism in Paris, but many of the Surrealist artists of the later decades also made use of it. Remedios Varo's œuvre could make a telling example.
30. Fore, *Realism after Modernism*, 6.
31. Paul Celan, *Selections*, ed. Pierre Joris (Berkeley: University of California Press, 2005), 158.
32. An interesting borderline case is the influence of the Scottish biologist D'Arcy Wentworth Thompson, whose "bio-mathematics" presented in his book *On Growth and Form* (Cambridge: Cambridge University Press, 2014) is known to have had a great impact on such different artists as Henry Moore, Richard Hamilton, Eduardo Paolozzi & Ben Nicholson. Also, many important modernist architects were impressed by the content of Thompson's book, László Moholy-Nagy and Ludwig Mies van der Rohe among them. (See "Sketching the Universe: Artists' Responses to D'Arcy Thompson," accessed November 17, 2016, www.dundee.ac.uk/museum/exhibitions/zoology/sketching/.)
33. With Greenbergian modernism I mean the tradition of American abstract art, which was backed by Clement Greenberg's theory that emphasized medium specificity and formalist aspects, and had its heyday after the Second World War. According to Greenberg's famous phrase, the essence of modernist painting was in his opinion "in the use of the characteristic methods of a discipline to criticize the discipline itself—not in order to subvert it, but to entrench it more firmly in its area of competence." Clement Greenberg, "Modernist Painting," in *The New Art: A Critical Anthology*, ed. Gregory Battock (New York: Dutton, 1973), 67.
34. Karen Kurczynski, *Art and Politics of Asger Jorn: The Avant-Garde Won't Give Up* (Burlington, VT: Ashgate, 2014), 86.
35. The animal at large played an important role in Beuys's art. Beuys is well known for his animal drawings (see for example Joseph Beuys, *Zeichnungen / Drawings*, ed. Heiner Bastian [Bielefeld: Kerber Verlag, 2014]), and one of the most renowned of his performances is about the artist, trying to solve the problem, as the name of the performance tells, "How to Explain Pictures to a Dead Hare."
36. In May 1974, the performance was held in the René Block Gallery at 409 West Broadway, New York, NY. See Caroline Tisdall, *Coyote* (Munich: Schirmer/Mosel, 2008).
37. Olly & Suzi are perhaps the most notable of the performance artists who have managed to go beyond the boundaries, with their art happenings and recordings in which the animals take part.
38. Claude Lévi-Strauss, *The Savage Mind*, trans. anonymous (Chicago: University of Chicago Press, 1970), 204.
39. Of the gesture, see for example Marcel Broodthaers, *Collected Writings,* ed. Gloria Moure (Barcelona: Ediciones Polígrafa, 2012), 100–117; Rachel Haidu, *The Absence of Work: Marcel Broodthaers 1964–1976* (Cambridge, MA: MIT Press, 2010), 46–106.
40. Duchamp, in Calvin Tomkins, *Bride and the Bachelors* (London: Penguin, 1968), 13. The casting into plaster of *Pense-bête* has been also seen as a Duchampian gesture of leaving one medium in favor of another: in Broodthaers's case, poetry for visual art; in Duchamp's, painting for conceptual art. See e.g. Haidu, *The Absence of Work*, 46–106.
41. See ibid., 19.
42. Broodthaers, cited in ibid., 207.
43. Broodthaers, cited in ibid., 167–168.
44. To consult the transcription of the recording, see Broodthaers, *Collected Writings*, 288–289.
45. E. T. A. Hoffmann, *The Life and Opinions of the Tomcat Murr*, trans. Anthea Bell (London: Penguin Books, 1999), 10–11.
46. Michel de Montaigne, *Apology for Raymond Sebond*, trans. M. A. Screech (Cambridge, MA: Hackett Publishing Company, 2003), 15.

Afterword

Thus, this is a history of the modern other. As contemporary critics have remarked, the history is then not animality discussed in the plural, but of a category more hybrid, more unstable, and more labelled by the human culture whose reverse it represents. This is perhaps more than ever pointed out in the moments when the practitioners of the theory tried to get rid of this quality; such is the lesson proved by the history that followed in Darwinism's wake, and that in one way or the other culminated in the black years of Surrealism; when the human-animal borderline was truly brought into question, the primitive, black forces behind the civilized habits were highlighted—the ethical other, once again.

Among my descriptions of the history of the concept of the animal in the modern age, I have tried to show some reasons why the animal took for itself the value described. This is of course a very large question, and it can be answered here only in light of what I have been calling the theory of the arts. In addition, the questions I have been asking are supposed to make the reader understand the *status* of the animal within the discussions of art in the modern age, that is, its structural position. My work does not call upon the truth about animality and animals at large, but tries to provide keys for the understanding of the history of the human fantasy about animality in the context discussed—a fantasy that was not without consequences.

Once more, the intention of the work is not to separate the modern from the postmodern too clearly, but to show from which premises the contemporary interest in picturing animals in art spaces, as well as the contemporary criticism towards the modern use and the modern attitude towards the animals that prevailed in the modernist practices sprang forth. I also hope to have shown that the attitude was more ambiguous and more manifold than it is usually said to be. This said, there are of course major differences, not the least among them the significant role that ecological questions have been playing in the field of the arts ever since Joseph Beuys's remarkable first initiatives. The animal has in them an evident place. Another is the interest that the philosophical criticism of art has aroused in what has come to be known as "becomings," and which has descended to its art theoretical usage from the paths opened by the works of Gilles Deleuze and Félix Guattari. The two writers have acquired a considerable status in the field with their idea about what they call "*becoming-animal*," a metamorphosis that can be momentarily made to appear for example, through the means of art[1]. However, it has a role as a mediator between the fields of what can be considered as modern and the postmodern in a remarkable way, since before affirming its unique role in the discussions of the last two or three decades of contemporary animal art—philosophical discussions that interpret the contemporary art field as more or less determined or permeated by Deleuze's and Guattari's philosophical investigation—one should remember that the concepts have on the contrary been formulated to make understandable their readings of such emblematically modern artists as Franz Kafka or Francis Bacon.[2]

However, what makes Deleuze's and Guattari's success perhaps more understandable among the theorists of the postmodern is the rupture to which the work of the two testifies. It is perhaps not their anti-modernity—whatever that is—that has the appeal, but the view that the metamorphosis, the "becoming-animal," has had a substantial role in our culture. The glance that they have been able to take—whatever its philosophical implications[3]—has given an assurance that contemporary animal art did not spring from nothingness, and that it was not, from the start, without a plausible philosophical basis. The Deleuzian framework has also managed at least in some sense to appropriate the "minor traditions"[4] of modernism with the Darwinist dream underlying Nietzsche's work: such a representation of a metamorphosis does not in the smallest extent make the conceptions of the culturally other invalid; on the contrary, it reinforces their value, but it can also be used to show the disproportion that the concept of the other establishes with regard to biological facts.[5] This has partly been able to make possible the fact that the idea of the complexity of the patterns of becomings, the quality that Deleuze's philosophy supports, has started to be inscribed into the discourse on the animal and the arts, a development still in process. Donna Haraway, Rosi Braidotti, and Elizabeth Grosz have, each on their own terms, given through their feminist theories, tools to continue on these paths, also in connection with the concept of the animal.[6] And evidently, the subsequent discourses have managed to go far beyond the restrictions of high modernism.

Finally, it is also understandable that at the time of such a movement as transgenic art, which is capable of *actually* putting the boundaries of the different species in question and makes the process a part of an artistic expression, the more or less existential way of putting the question on the divide has still preserved its allure, also in the theories just cited—and it has perhaps come to the fore more strongly than ever. This is one of the lessons to be learned from Jacques Derrida's late works that also had an effect on the field. As such, the animal has mostly stayed in its place, representing to the human gaze, in Carsten Höller's sublime formulation, "a dysfunctional mirror in the frustrating quest of trying to see ourselves,"[7] and from this it has derived its distinguished place in the field of visual art. In other words, the animal still provides for us an image of a world that—whether for fantasy, for becomings, for empathy, for imagination, for the unknown, or for ill, or for worse—can be made to represent what constantly sleeps in us.

Notes

1. See Gilles Deleuze & Félix Guattari, *A Thousand Plateaus: Capitalism and Schizophrenia*, trans. Brian Massumi (London: Continuum, 2004), 256–341.
2. Most notably in connection with Franz Kafka; see Gilles Deleuze & Félix Guattari, *Kafka: Toward a Minor Literature*, trans. Dana Polan (Minneapolis: University of Minnesota Press 1986). In connection with modernist painting and Francis Bacon: Gilles Deleuze, *Francis Bacon: The Logic of Sensation*, trans. Daniel W. Smith (London: Continuum, 2003), 20–26.

3. In *Thousand Plateaus*, Deleuze and Guattari speak of the "becoming-animal" in its particular connections with their idea of the "rhizome." See Deleuze & Guattari, *A Thousand Plateaus*, 263.
4. For Deleuze's conception of "minor" tradition in connection with his idea of the "becoming-animal," see Deleuze & Guattari, *Kafka*.
5. A strange political idealism is hidden in Deleuze's and Guattari's idea, through which they ambiguously validate their connections with the concept of the other and of the minority, while at the same time renewing their critique of representations by enforcing, among other dualities, the human-animal divide. They write, "Moreover, what Messiaen discovered in music is the same thing that the ethologists discovered in animals: human beings are hardly at an advantage, except in the means of overcoding, of making punctual systems. That is even the opposite of having an advantage: through becomings-woman, -child, -animal, or -molecular, nature opposes its power, and the power of music, to the machines of human beings, the roar of factories and bombers." (Deleuze & Guattari, *A Thousand Plateaus*, 341.) My aim is not here to criticize the political implications or the philosophical coherence of these arguments, but to state that their idea, as epoch-making as it may be in some sense, is a continuation of the modernist thinking on the subject.
6. In connection with our topic, see for example Rosi Braidotti, *Metamorphoses: Towards a Materialist Theory of Becoming* (Malden, MA: Polity, 2002); Elizabeth Grosz, *Chaos, Territory, Art: Deleuze and the Framing of the Earth* (New York: Columbia University Press, 2012), especially 63–66.
7. Höller, cited in Giovanni Aloi, *Art and Animals* (London: I.B. Tauris, 2011), 17.

Bibliography

Agamben, Giorgio. *The Open: Man and Animal.* Translated by Kevin Attell. Stanford, CA: Stanford University Press, 2004.

Aloi, Giovanni. *Art and Animals.* London: I.B. Tauris, 2011.

Ansell-Pearson, Keith. *Viroid Life: Perspectives on Nietzsche and the Transhuman Condition.* New York: Routledge, 1997.

Apollinaire, Guillaume. "Pure Painting." In *The Modern Tradition: Backgrounds of Modern Literature*, edited by Richard Ellmann & Charles Feidelson, Jr., 113–115. New York: Oxford University Press, 1965.

Aristotle. *The Complete Works of Aristotle.* Edited by Jonathan Barnes. Princeton: Princeton University Press, 1995.

Avital, Tsion. "The Origins of Art: An Archaeological or a Philosophical Problem?" *South African Journal of Art History*, vol.16 (2001): 34–57.

Baker, Steve. *Picturing the Beast: Animals, Identity and Representation.* Champaign: University of Illinois Press, 2001.

Baker, Steve. *The Postmodern Animal.* London: Reaktion Books, 2000.

Barnes, Sara & Andrew Patrizio. "Darwin on the Threshold of the Visible: Contemporary Art and Evolution." In *The Art of Evolution: Darwin, Darwinisms, and Visual Culture*, edited by Barbara Larson & Fae Brauer, 288–312. Hanover, NH: Dartmouth College Press, 2009.

Barr, Alfred H. Jr. *Cubism and Abstract Art.* New York: Museum of Modern Art, 1936.

Bataille, Georges. *Cradle of Humanity: Prehistoric Art & Culture.* Edited by Stuart Kendall. Translated by Michelle Kendall & Stuart Kendall. New York: Zone Books, 2009.

Bataille, Georges. *Œuvres complètes,* tomes I–XII. Paris: Gallimard, 1971–1988.

Bataille, Georges. *Theory of Religion.* Translated by Robert Hurley. New York: Zone Books, 1994.

Batteux, Charles. "The Fine Arts Reduced to a Single Principle." In *The Bloomsbury Anthology of Aesthetics,* edited by Joseph Tanke & Colin McQuillan, 140–157. New York: Bloomsbury, 2012.

Baudelaire, Charles. *Écrits sur l'art.* Paris: Le livre de poche, 1992.

Baumgarten, Alexander Gottlieb. "Aesthetica." In *The Bloomsbury Anthology of Aesthetics,* edited by Joseph Tanke & Colin McQuillan, 158–162. New York: Bloomsbury, 2012.

Bentham, Jeremy. *An Introduction to the Principles of Morals and Legislation.* London: Methuen, 1982.

Berri, Kenneth. "Diderot's Hieroglyphs." *SubStance, Issue* 92 29, no. 2 (2000): 68–93.

Beuys, Joseph. *Zeichnungen / Drawings.* Edited by Heiner Bastian. Bielefeld: Kerber Verlag, 2014.

Blake, William. *The Marriage of Heaven and Hell.* Adelaide: University of Adelaide Library 2004. Accessed February 4, 2015. https://ebooks.adelaide.edu.au/b/blake/william/marriage/.

Blavatsky, H. P. *Isis Unveiled: a Master-Key to the Mysteries of Ancient and Modern Science and Theology*. Point Loma, CA: The Theosophical Publishing Company, 1906.

Boucher, Armand. *Darwinisme et socialisme*. London: Forgotten Books, 2013. Accessed August 20, 2014. www.forgottenbooks.com/readbook_text/Darwinisme_et_Socialisme_1200093082/45.

Braidotti, Rosi. *Metamorphoses: Towards a Materialist Theory of Becoming*. Malden, MA: Polity, 2002.

Brain, Robert Michael. "Protoplasmania: Huxley, Haeckel, and the Vibratory Organism in Late Nineteenth-Century Science and Art." In *The Art of Evolution: Darwin, Darwinisms, and Visual Culture*, edited by Barbara Larson & Fae Brauer, 92–123. Hanover, NH: Dartmouth College Press, 2009.

Braudel, Fernand. *A History of Civilizations*. Translated by Richard Mayne. London: Penguin Books, 1993.

Brauer, Fae. "Wild Beasts and Tame Primates: 'Le Douanier' Rousseau's Dream of Darwin's Evolution." In *The Art of Evolution: Darwin, Darwinisms, and Visual Culture*, edited by Barbara Larson & Fae Brauer, 194–225. Hanover, NH: Dartmouth College Press, 2009.

Breton, André. "Fiches de préparation à l'Exposition." Accessed November 19, 2016. www.andrebreton.fr/work/56600100071620

Breton, André. *Manifestoes of Surrealism*. Translated by Richard Seaver & Helen R. Lane. Ann Arbor, MI: The University of Michigan Press, 1969.

Breuil, Henri. *Four Hundred Centuries of Cave Art*. Translated by Mary E. Boyle. New York: Hacker Art Books, 1979.

Breuil, Henri. "Paleolithic Art (art mobilier)." *Fonds Breuil, Br 48.8*. Muséum national d'Histoire naturelle, Paris.

Breuil, Henri & Émile Cartailhac. *La Caverne d'Altamira à Santillane près Santander (Espagne)*. Monaco: Imprimerie de Monaco, 1906.

Broodthaers, Marcel. *Collected Writings*. Edited by Gloria Moure. Barcelona: Ediciones Polígrafa, 2010.

Bryson, Norman. *Word and Image: French Painting of the Ancien Régime*. Cambridge: Cambridge University Press, 1981.

Buñuel, Luis. *My Last Breath*. Translated by Abigail Israel. Glasgow: Fontana Paperbacks, 1985.

Bürger, Peter. *Theory of the Avant-Garde*. Translated by Michael Shaw. Minneapolis, MN: University of Minnesota Press, 1989.

Burke, Edmund. *A Philosophical Enquiry into the Origin of our Ideas of the Sublime and Beautiful*. Oxford: Oxford University Press, 1990.

Cabanne, Pierre. *Entretiens avec Marcel Duchamp*. Paris: Belfond, 1967.

Caillois, Roger. *The Edge of Surrealism: A Roger Caillois Reader*. Edited by Claudine Frank. Translated by Camille Naish. London: Duke University Press, 2003.

Cartailhac, Émile. "La grotte d'Altamira, Espagne. Mea culpa d'un sceptique." *L'Anthropologie* 13 (1902): 348–354.

Cassirer, Ernst. *An Essay on Man: An Introduction to the Philosophy of Human Culture*. Garden City, NY: Yale University Press, 1944.

Celan, Paul. *Selections*. Edited by Pierre Joris. Berkeley: University of California Press, 2005.

Cheung, Tobias. "Omnis Fibra ex Fibra: Fibre Economies in Bonnet's and Diderot's Models of Organic Order." In *Transitions and Borders Between Animals, Humans and Machines 1600–1800*, edited by Tobias Cheung. Leiden: Brill, 2010.

Chirico, Giorgio de. "Mystery and Creation." In *Art in Theory, 1900–2000: An Anthology of Changing Ideas*, edited by Charles Harrison & Paul Wood, 58. Oxford: Blackwell Publishing, 2003.

Clair, Jean, ed. *The 1930s: The Making of the New Man*. Ottawa: National Gallery of Canada, 2008.

Clark, Kenneth. *Animals and Men: Their Relationship as Reflected in Western Art from Prehistory to the Present Day*. London: Thames and Hudson, 1977.

Clark, T.J. *The Painting of Modern Life: Paris in the Art of Manet and His Followers*. London: Thames and Hudson, 1990.

Clottes, Jean & David Lewis-Williams. *The Shamans of Prehistory: Trance and Magic in the Painted Caves*. New York: Harry N. Abrams, 1998.

Cohen, Sarah R. "Animal Performance in Oudry's Illustrations to the Fables of La Fontaine." *Studies in Eighteenth Century Culture* 39 (2010): 35–76.

Condillac, Étienne Bonnot de. *Essay on the Origin of Human Knowledge*. Translated by Hans Aarsleff. Cambridge: Cambridge University Press, 2001.

Condillac, Étienne Bonnot de. *Traité des animaux*. Paris: J. Vrin, 2004.

Connelly, Frances S. *The Sleep of Reason: Modern European Art and Aesthetics, 1725–1907*. University Park: The Pennsylvania State University Press, 1995.

Dagognet, François. *L'animal selon Condillac: Étude sur le "Traité des animaux."* Paris: J.Vrin, 2004.

Dalí, Salvador. "L'âne pourri." In "*Il y aura une fois": Une anthologie du Surréalisme*. Edited by Jacqueline Chénieux-Gendron, 183–186. Paris: Gallimard, 2003.

Danesi, Marcel. *Vico, Metaphor, and the Origin of Language*. Bloomington: Indiana University Press, 1993.

D'Arcy Wentworth Thompson. *On Growth and Form*. Cambridge: Cambridge University Press, 2014.

Darwin, Charles. *The Descent of Man and Selection in Relation to Sex*. London: Murray, 1899.

Darwin, Charles. *Expression of the Emotions in Man and Animals*. London: HarperCollins, 1998.

Darwin, Charles. *On the Origin of Species*. Calgary: Broadview Press, 2003.

Daston, Lorraine J. & Peter Galison. *Objectivity*. Cambridge, MA: Zone Books, 2010.

Delacroix, Eugène. *Journal, 1822–1863*. Paris: Plon, 1981.

Deleuze, Gilles. *Francis Bacon: The Logic of Sensation*. Translated by Daniel W. Smith. London: Continuum, 2003.

Deleuze, Gilles & Félix Guattari. *A Thousand Plateaus: Capitalism and Schizophrenia*. Translated by Brian Massumi. London: Continuum, 2004.

Deleuze, Gilles & Félix Guattari. *Kafka: Toward a Minor Literature*. Translated by Dana Polan. Minneapolis: University of Minnesota Press, 1986.

Derrida, Jacques. *The Animal That Therefore I Am*. Translated by David Wills. New York: Fordham University Press, 2008.

Derrida, Jacques. *The Beast & the Sovereign: Volume I*. Translated by Geoffrey Bennington. Chicago: The University of Chicago Press, 2009.

Derrida, Jacques. *The Beast & the Sovereign: Volume II*. Translated by Geoffrey Bennington. Chicago: The University of Chicago Press, 2010.

Derrida, Jacques. *Of Grammatology*. Translated by Gayatri Chakravorty Spivak. London: Johns Hopkins University Press, 1976.

Descartes, René. *Oeuvres de Descartes*, 11 Vols. Paris: J. Vrin, 1996.

Descartes, René. *Philosophical Works of Descartes*, Vol III. Edited by John Cottingham, Robert Stoothoff, Dugald Murdoch & Anthony Kenny. Cambridge: Cambridge University Press, 1991.

Diderot, Denis, *A Conversation between D'Alembert and Diderot*. Translated by Ian Johnston. Nanaimo, BC: Johnstonia. Accessed January 2, 2015. www.oneeyedman.net/school-archive/classes/fulltext/www.mala.bc.ca/~johnstoi/diderot/conversation.htm.

Diderot, Denis. *Rêve d'Alembert*. Translated by Ian Johnston. Nanaimo, BC: Johnstonia. Accessed January 2, 2015. http://records.viu.ca/~johnstoi/diderot/revedalembert_tofc.htm.

Diderot, Denis. *Salons*. Edited by Michel Delon. Paris: Gallimard, 2008.

Diderot, Denis. *Salons*. Vol. 3. Edited by Jean Seznac & Jean Adhémar. London: Clarendon Press, 1979.

Diderot, Denis. *Salons*. Vol. 4. Edited by Jean Seznac & Jean Adhémar. London: Clarendon Press, 1979.

Dottori, Riccardo. "The Metaphysical Parable in Giorgio de Chirico's Painting." *Metafisica*, no. 5–6 (2006): 203–213.

Duchamp, Marcel. *Affectt Marcel: The Selected Correspondence of Marcel Duchamp*. Edited by Francis M. Naumann & Hector Obalk. London: Thames and Hudson, 2000.

Duchamp, Marcel. *The Writings of Marcel Duchamp*. Edited by Michel Sanouillet & Elmer Peterson. Oxford: Oxford University Press, 1973.

Eisenman Stephen F. *The Cry of Nature: Art and the Making of Animal Rights*. London: Reaktion Books, 2013.

Éluard, Paul. *Œuvres complètes*, tome 1. Paris: Gallimard, 1993.

Ernst, Max. "An Informal Life of M.E. (as told by himself to a young friend)." In *Max Ernst*, edited by William S. Liebermann, 7–24. New York: The Museum of Modern Art, 1961.

Fabre, J.H. *Social Life in the Insect World*. Translated by Bernard Miall. Accessed February 12, 2013. www.gutenberg.org/files/18350/18350-h/18350-h.htm.

Fernow, Karl Ludwig. "On Landscape Painting." In *Art in Theory, 1648–1815: An Anthology of Changing Ideas*, edited by Charles Harrison, Paul Wood & Jason Gaiger. Translated by Jason Gaiger, 1066–1071. Oxford: Blackwell Publishing, 2000.

Fiedler, Conrad. *On Judging Works of Visual Art*. Translated by Henry Schaefer-Simmern & Fulmer Mood. Berkeley: University of California Press, 1978.

Fischer, Luke. "Animalising Art: Rainer Maria Rilke and Franz Marc." *Australasian Journal of Ecocriticism and Cultural Ecology* 3 (2013/2014): 45–60.

Fore, Devin. *Realism After Modernism: The Rehumanization of Art and Literature*. Cambridge, MA: MIT Press, 2012.

Foster, Hal. *Compulsive Beauty*. Cambridge, MA: MIT Press, 1993.

Foucault, Michel. *Madness and Civilization: A History of Insanity in the Age of Reason*. Translated by Richard Howard. London: Routledge, 2009.

Foucault, Michel. *Order of Things: An Archaeology of the Human Sciences*. London: Continuum, 2002.

Frank, Adolphe, ed. *Dictionnaire des sciences philosophiques*. Vol. 2. Paris: Hachette, 1844–1852.

Frazer, James George. *The Golden Bough: A Study in Magic and Religion*. Oxford: Oxford University Press, 2009.

Freud, Sigmund. "Beyond the Pleasure Principle." In *The Standard Edition of the Complete Psychological Works of Sigmund Freud*. Vol. 18, edited by James Strachey, 7–66. London: Hogarth Press, 1956–1974.

Freud, Sigmund. "Civilization and Its Discontents." In *The Standard Edition of the Complete Psychological Works of Sigmund Freud*. Vol. 21, edited by James Strachey, 59–148. London: Hogarth Press, 1956–1974.

Fried, Michael. *Absorption and Theatricality: Painting and Beholder in the Age of Diderot*. Chicago: Chicago University Press, 1980.

Fried, Michael. *Courbet's Realism*. Chicago: The University of Chicago Press, 1990.

Fried, Michael. *Manet's Modernism, or the Face of Painting in the 1860s*. Chicago: Chicago University Press, 1996.

Fried, Michael. *The Moment of Caravaggio*. Princeton, NJ: Princeton University Press, 2010.

Fry, Roger. "An Essay in Aesthetics." In *Art in Theory, 1900–2000: An Anthology of Changing Ideas*, edited by Charles Harrison & Paul Wood, 75–82. Oxford: Blackwell Publishing, 2003.

Funt, David. "Diderot and the Esthetics of Enlightenment." In *Diderot Studies XI*, edited by Otis Fellows & Diana Guiragossian, 463–483. Genève: Librairie Droz, 1968.

Fuseli, Henry. "Introduction to *The Grave*." In *Art in Theory, 1648–1815: An Anthology of Changing Ideas*, edited by Charles Harrison, Paul Wood & Jason Gaiger, 1008–1010. Oxford: Blackwell Publishing, 2000.

Gautier, Théophile. "From Preface to *Mademoiselle de Maupin*." In *Art in Theory, 1815–1900: An Anthology of Changing Ideas*, edited by Charles Harrison, Paul Wood, and Jason Gaiger, 96–100. Oxford: Blackwell Publishing, 1998.

Gay, Peter. *Weimar Culture: The Outsider as Insider*. New York: Harper & Row, 2001.

Goethe, Johann Wolfgang von. *Goethe on Art*. Edited by John Gage. Berkeley: University of California Press, 1980.

Goethe, Johann Wolfgang von. "Maxims and Reflections." In *Art in Theory, 1815–1900: An Anthology of Changing Ideas*, edited by Charles Harrison, Paul Wood & Jason Gaiger. Translated by Nicholas Walker. 74–78. Oxford: Blackwell Publishing, 1998.

Gombrich, E. H. *The Preference for the Primitive: Episodes in the History of Western Taste and Art*. London: Phaidon, 2002.

Goncourt, Edmond & Jules Goncourt. *Études d'art: Le Salon de 1852—La peinture à l'exposition 1855*. Paris: Librairie des Bibliophiles. Accessed December 18, 2016. https://archive.org/stream/gri_33125008506780#page/n3/mode/2up.

Gott, Todd. "Clutch of the Beast: Emmanuel Frémiet, Gorilla-Sculptor." In *Kiss of the Beast: From Paris Salon to King Kong*. Brisbane: Queensland Art Gallery, 2005.

Greenberg, Clement. "Modernist Painting." In *The New Art: A Critical Anthology*, edited by Gregory Battock, 67. New York: Dutton, 1973.

Grén, Roni. *Georges Bataille, historia ja taide*. Turku: Turun yliopisto, 2012.

Grén, Roni. *"Luolamaalaukset taiteena—uskonnollisen tulkinnan variaatioita Salomon Reinachilla, Abbé Breuililla ja Georges Bataillella."* In *Tahiti—taidehistoria tieteenä* (January 2015). Accessed November 17, 2016. http://tahiti.fi/01–2015/tieteelliset-artikkelit/luolamaalaukset-taiteena—-uskonnollisen-tulkinnan-variaatioita-salomon-reinachilla-abbe-breuililla-ja-georges-bataillella/.

Grosz, Elizabeth. *Chaos, Territory, Art: Deleuze and the Framing of the Earth*. New York: Columbia University Press, 2008.

Gruber, Howard E. *Darwin on Man: A Psychological Study of Scientific Creativity, together with Darwin's Early and Unpublished Notebooks*. Chicago: University of Chicago Press, 1981.

Guichet, Jean-Luc. *Enjeux de la question de l'animal sous les Lumières: Condillac, Diderot, Rousseau*. Accessed January 10, 2015. http://ecole-thema.ens-lyon.fr/IMG/pdf/Article_Guichet-2.pdf.

Haeckel, Ernst. *The History of Creation: Or the Development of the Earth and Its Inhabitants by the Action of Natural Causes*. Translated by Edwin Ray. Accessed November 17, 2016. https://archive.org/details/historycreation00lankgoog.

Haeckel, Ernst. *Natürliche Schöpfungs-Geschichte: Gemeinverständliche wissenschaftliche vorträge über die entwickelungslehre im allgemeinen und diejenige von Darwin, Goethe und Lamarck im besonderen*. Berlin: Druck und Verlag von Georg Reimer, 1889. Accessed March 20, 2017. Available from: https://archive.org/details/natrlichesch00haec.

Haeckel, Ernst. "Zellseelen und Seelenzellen." Leipzig: Alfred Kröner Verlag, 1909. Accessed November 17, 2016. www.zum.de/stueber/haeckel/zellseelen/index.html.

Haidu, Rachel. *The Absence of Work: Marcel Broodthaers 1964–1976*. Cambridge, MA: MIT Press, 2010.

Hegel, Georg Wilhelm Friedrich. *Introductory Lectures on Aesthetics*. Translated by Bernard Bosanquet. London: Penguin Books, 2004.

Hegel, Georg Wilhelm Friedrich. *Philosophy of History*. Translated by J. Sibree. Marxists Internet Archive. Accessed May 15, 2014. www.marxists.org/reference/archive/hegel/works/hi/index.htm.

Heidegger, Martin. *The Basic Writings*. Edited by David Farrell Krell. New York: HarperCollins, 2008.

Heidegger, Martin. *Being and Time*. Translated by Joan Stambaugh. New York: State University of New York Press, 1996.

Heidegger, Martin. *The Fundamental Concepts of Metaphysics: World, Finitude, Solitude*. Translated by William McNeill & Nicholas Walker. Indianapolis: Indiana University Press. 1995

Henry, Victor. *La magie dans l'Inde antique*. Paris: Ernest Leroux, 1904.

Herder, Johann Gottfried von. "Essay on the Origin of Language." *On the Origin of Language*. Translated by John H. Moran & Alexander Gode, 85–176. Chicago: University of Chicago Press, 1986.

Heymans, Peter. "An Ecocritical Approach to the Discourse of the Sublime in 'The Ancient Mariner'." *The Coleridge Bulletin: The Journal of the Friends of Coleridge New Series* 30 (Winter 2007): 17–26.

Hirn, Yrjö. *The Origins of Art: A Psychological and Sociological Inquiry*. London: Macmillan, 1900.

Hoffmann, E.T.A. *The Life and Opinions of the Tomcat Murr*. Translated by Anthea Bell. London: Penguin Books, 1999.

Husserl, Edmund. *Collected Works, Volume Three: Ideas Pertaining to a Pure Phenomenology and to a Phenomenological Philosophy*. Translated by Richard Rojcewicz & André Schuwer. London: Kluwer Academic Publishers. Accessed May 15, 2014. https://ia801402.us.archive.org/20/items/IdeasPartIi/Husserl-IdeasIi.pdf.

Hutcheson, Francis. "An Inquiry into the Original of Our Ideas of Beauty and Virtue." In *The Bloomsbury Anthology of Aesthetics*, edited by Joseph Tanke & Colin McQuillan, 124–132. New York: Bloomsbury, 2012.

Itten, Johannes. "Analysis of the Old Masters." In *Art in Theory, 1900–2000: An Anthology of Changing Ideas*, edited by Charles Harrison & Paul Wood, 304–306. Oxford: Blackwell Publishing, 2003.

Jay, Martin. *Downcast Eyes: The Denigration of Vision in Twentieth-Century French Thought*. Berkeley: University of California Press, 1993.

Kalof, Linda. *Looking at Animals in Human History*. London: Reaktion Books, 2007.

Kandinsky, Wassily. "The Cologne Lecture." In *Art in Theory, 1900–2000: An Anthology of Changing Ideas*, edited by Charles Harrison & Paul Wood, 89–93. Oxford: Blackwell Publishing, 2003.

Kandinsky, Wassily. *Concerning the Spiritual in Art*. Translated by Michael T.H. Sadler. Accessed March 21, 2015. www.semantikon.com/art/kandinskyspiritualinart.pdf.

Kandinsky, Wassily. *Point and Line to Plane: Contribution to the Analysis of the Pictorial Elements*. Translated by Howard Dearstyne & Hilla Rebay. New York: Solomon R. Guggenheim Foundation, 1947. Accessed November 20, 2016. https://archive.org/details/pointlinetoplane00kand.

Kant, Immanuel. *Critique of Judgment*. Translated by James Creed Meredith. Mineola, NY: Dover, 2012.

Kant, Immanuel. *Critique of the Power of Judgment*. Edited by Paul Guyer. Cambridge: Cambridge University Press, 2000.

Kant, Immanuel. *The Critique of Pure Reason*. Translated by J.M.D. Meiklejohn. Salt Lake City, UT: Project Gutenberg. Accessed May 15, 2014. www.gutenberg.org/files/4280/4280-h/4280-h.htm.

Kant, Immanuel. *Lectures on Metaphysics*. Edited by Karl Ameriks & Steve Naragon. Cambridge: Cambridge University Press, 1997.

Kemp, Martin. *Human Animal in Western Art and Science*. Chicago: University of Chicago Press, 2007.

Kierkegaard, Søren. *Kierkegaard's Writings IV, Part II: Either/Or*. Translated by Howard W. Hong & Edna H. Hong. Princeton, NJ: Princeton University Press, 2013.

Kirchner, Horst. "Ein Beitrag zur Urgeschichte der Schamanismus." *Anthropos* 47, no. 1/2 (1952): 244–286.

Klee, Paul. "Ways of Studying Nature." In *Paul Klee: Dialogue with Nature*, edited by Ernst-Gerhard Güse, 8. Munich: Prestel, 1991.

Knight, Richard Payne. *An Inquiry into the Symbolical Language of Ancient Art and Mythology*. New York: J.W. Bouton, 1892.

Kojève, Alexandre. *Introduction to the Reading of Hegel*. Translated by James H. Nichols, Jr. Ithaca, NY: Cornell University Press, 1980.

Krantz, Émile. *L'èsthétique de Descartes: rapports de la doctrine cartésienne avec la littérature classique française au XVIIe siècle*. Paris: Félix Alcan, 1902. Accessed January 6, 2016. https://ia600408.us.archive.org/13/items/essaisurlestht00kran/essaisurlestht00kran.pdf

Kristeva, Julia. *Powers of Horror: An Essay on Abjection*. Translated by Leon S. Roudiez. New York: Columbia University Press, 1982.

Kurczynski, Karen. *Art and Politics of Asger Jorn: The Avant-Garde Won't Give Up*. Burlington, VT: Ashgate, 2014.

Lacan, Jacques. *Écrits: A Selection*. Translated by Alan Sheridan. London: Tavistock, 1985.

Lacan, Jacques. *The Four Fundamental Concepts of Psycho-Analysis*. Translated by Alan Sheridan. London: Karnac, 1977.

LaCapra, Dominick. *History and Its Limits: Human, Animal, Violence*. London: Cornell University Press, 2009.

La Font de Saint-Yenne, Étienne. "Réflexions sur quelques causes de l'état présent de la peinture en France, avec un examen de principaux ouvrages exposés au Louvre." In *Œuvres critiques*, edited by Étienne Jollet. Paris: ENSBA, 2001.

La Mettrie, Julian Offray de. *Machine Man and Other Writings*. Edited by Ann Thomson. Cambridge: Cambridge University Press, 1996.

Langer, Monika. "The Role and the Status of the Animals in Nietzsche's Philosophy." In *Animal Others: On Ethics, Ontology and Animal Life*, edited by H. Peter Steeves, 75–92. New York: State University of New York Press, 1999.

Larson, Barbara. "Darwin's Sexual Selection and the Jealous Male." In *The Art of Evolution: Darwin, Darwinisms, and Visual Culture*, edited by Barbara Larson & Fae Brauer, 173–193. Hanover, NH: Dartmouth College Press, 2009.

Larson, Barbara. "Introduction." In *The Art of Evolution: Darwin, Darwinisms, and Visual Culture*, edited by Barbara Larson & Fae Brauer, 1–17. Hanover, NH: Dartmouth College Press, 2009.

Le Dantec, Félix. *Influences ancestrales*. Paris: Flammarion, 1920.

Leibniz, Gottfried Wilhelm. *Philosophical Papers and Letters*. Edited by Leroy E. Loemker. Dordrecht: Reidel, 1969.

Leibniz, Gottfried Wilhelm. *Philosophical Writings*. Edited by G.H.R. Parkinson. London: Dent, 1973.

Leibniz, Gottfried Wilhelm. *Sämtliche Schriften und Briefe*. Vol. 6. Berlin: Akademie Verlag, 1923.

Lemm, Vanessa. *Nietzsche's Animal Philosophy: Culture, Politics, and the Animality of the Human Being*. New York: Fordham University Press, 2009.

Lessing, Gotthold Ephraim. *Laocoon: An Essay on the Limits of Painting and Poetry*. Translated by Edward Allen McCormick. Baltimore, MD: Johns Hopkins University Press, 1984.

Lévi-Strauss, Claude. *The Savage Mind*. Translator anonymous. Chicago: University of Chicago Press, 1970.

Lewis-Williams, David. *The Mind in the Cave: Consciousness and the Origins of Art*. London: Thames and Hudson, 2002.

Lichtenstein, Jacqueline. *Les raisons de l'art. Essais sur les limites de l'esthétique*. Paris: Gallimard, 2014.

Lista, Marcella. "Between Heaven and Earth: The Abstract Utopia of Franz Marc and the Artistic Theory of His Time." *Chicago Art Journal* 10 (2000): 1–24.

Locke, John. *An Essay Concerning Human Understanding*, Volume I. Accessed May 15, 2014. www.gutenberg.org/cache/epub/10615/pg10615.html.

Maciunas, George. "Neo-Dada in Music, Theater, Poetry, Art." In *Art in Theory, 1900–2000: An Anthology of Changing Ideas*, edited by Charles Harrison & Paul Wood, 727–729. Oxford: Blackwell Publishing, 2003.

Macke, August. "Masks." In *Art in Theory, 1900–2000: An Anthology of Changing Ideas*, edited by Charles Harrison & Paul Wood, 95–96. Oxford: Blackwell Publishing, 2003.

McLear, Colin. "Kant on Animal Consciousness." *Philosopher's Imprint* 11, no. 15 (November 2011): 1–16.

Malevich, Kasimir. "From Cubism and Futurism to Suprematism: The New Realism in Painting." In *Art in Theory, 1900–2000: An Anthology of Changing Ideas*, edited by Charles Harrison & Paul Wood, 173–183. Oxford: Blackwell Publishing, 2003.

Malevich, Kasimir. "The Question of Imitative Art." In *Art in Theory, 1900–2000: An Anthology of Changing Ideas*, edited by Charles Harrison & Paul Wood, 293–298. Oxford: Blackwell Publishing, 2003.

Malraux, André. *The Voices of Silence*. Translated by Stuart Gilbert. Princeton, NJ: Princeton University Press, 1978.

Marc, Franz. "How Does a Horse See the World." In *Theories of Modern Art: A Source Book by Artists and Critics*, edited by Herschel B. Chipp, Peter Selz & Joshua C. Taylor. Berkeley: University of California Press, 1968.

Marinetti, F. T. "The Founding and Manifesto of Futurism." In *Futurism: An Anthology*, edited by Lawrence Rainey, Christine Poggi & Laura Wittman, 49–53. New Haven, CT: Yale University Press, 2009.

Marx, Karl. *Economic & Philosophic Manuscripts of 1844*. Translated by Martin Milligan, revised by Dirk J. Struik. Marxists Internet Archive. Accessed May 15, 2014. www.marxists.org/archive/marx/works/download/pdf/Economic-Philosophic-Manuscripts-1844.pdf.

Mathews, Patricia Townley. *Passionate Discontent: Creativity, Gender, and French Symbolist Art*. Chicago: University of Chicago Press, 1999.

Merleau-Ponty, Maurice. *La Nature: Notes du Collège de France*. Paris: Éditions du Seuil, 1994.

Mesch, Claudia, ed. *Joseph Beuys: The Reader*. London: I.B. Tauris, 2007.

Miles, Murray. "Leibniz on Apperception and Animal Souls." *Dialogue* 33 (Fall 1994): 701–724.

Mondrian, Piet. "Plastic Art and Pure Plastic Art." In *Art in Theory, 1900–2000: An Anthology of Changing Ideas*, edited by Charles Harrison & Paul Wood. Oxford: Blackwell Publishing, 2003.

Montaigne, Michel de. *Apology for Raymond Sebond*. Translated by M. A. Screech. Cambridge, MA: Hackett Publishing Company, 2003.

Morris, Katherine. "Bêtes-machines." In *Descartes' Natural Philosophy*, edited by Stephen Gaukroger, John A. Schuster, & John Sutton, 401–419. London: Routledge, 2000.

Morton, Marsha. "From *Monera* to Man: Ernst Haeckel, *Darwinismus*, and Nineteenth-Century German Art." In *The Art of Evolution: Darwin, Darwinisms, and Visual Culture*, edited by Barbara Larson & Fae Braue, 59–91. Hanover, NH: Dartmouth College Press, 2009.

Munch, Edvard. "Notebook and Diary Entries." In *Art in Theory, 1815–1900: An Anthology of Changing Ideas*, edited by Charles Harrison, Paul Wood & Jason Gaiger, 1039–1044. Oxford: Blackwell Publishing, 2000.

Newman, Barnett. "The First Man Was an Artist." In *Art in Theory, 1900–2000: An Anthology of Changing Ideas*, edited by Charles Harrison & Paul Wood, 574–577. Oxford: Blackwell Publishing, 2003.

Nietzsche, Friedrich. *The Antichrist, Ecce Homo, Twilight of the Idols, and Other Writings*. Translated by Judith Norman. Cambridge: Cambridge University Press, 2005.

Nietzsche, Friedrich. *Beyond Good and Evil: Prelude to a Philosophy of the Future*. Translated by Judith Norman. Cambridge: Cambridge University Press, 2002.

Nietzsche, Friedrich. *The Birth of Tragedy and Other Writings*. Translated by Ronald Speirs. Cambridge: Cambridge University Press, 1999.

Nietzsche, Friedrich. *Daybreak: Thoughts on the Prejudices of Morality*. Translated by R.J. Hollingdale. Cambridge: Cambridge University Press, 1997.

Nietzsche, Friedrich. *Digitale Kritische Gesamtausgabe Werke und Briefe*. Edited by Paolo d'Iorio. Berlin and New York: deGruyter, 1967. Accessed March 23 2015. www.nietzsche-source.org/#eKGWB.

Nietzsche, Friedrich. *The Gay Science: With a Prelude in German Rhymes and an Appendix of Songs*. Edited by Bernard Williams. Translated by Josefine Nauckhoff & Adrian del Caro. Cambridge: Cambridge University Press, 2001.

Nietzsche, Friedrich. *Thus Spoke Zarathustra*. Translated by Thomas Common. Paris: Feedbooks, 2010.

Nietzsche, Friedrich. *Untimely Meditations*. Translated by R.J. Hollingdale. Cambridge: Cambridge University Press, 1997.

Nietzsche, Friedrich. *The Will to Power*. Translated by Walter Kauffman & R.J. Hollingdale. New York: Vintage Books, 1967.

Nordau, Max. 1895. *Degeneration*. London: William Heinemann. Accessed December 7, 2014. https://archive.org/details/degeneration035137mbp.

Norris, Margot. *Beasts of the Modern Imagination*. Baltimore, MD: Johns Hopkins University Press, 1985.

Novalis. "Fugitive Thoughts." In *Art in Theory, 1648–1815: An Anthology of Changing Ideas*, edited by Charles Harrison, Paul Wood & Jason Gaiger, 909–913. Oxford: Blackwell Publishing, 2000.

Ottinger, Didier. *Surréalisme et mythologie moderne: les voies du labyrinthe d'Ariane à Fantômas*. Paris: Gallimard, 2002.

Palacio-Pérez, Eduardo. "Cave Art and the Theory of Art: The Origins of the Religious Interpretation of Palaeolithic Graphic Expression." *Oxford Journal of Archaeology* 29 (2010): 1–14.

Palacio-Pérez, Eduardo. "The Origins of the Concept of 'Palaeolithic Art': Theoretical Roots of an Idea." *Journal of Archaeological Method and Theory* 20 (December 2013): 682–714.

Palacio-Pérez, Eduardo. "Science and Belief in the Construction of the Concept of Paleolithic Religion." *Complutum* 24 (December 2013): 51–61.

Paradis, James. "Darwin and Landscape." In *Victorian Science and Victorian Values: Literary Perspectives*, edited by James Paradis & Thomas Postlewait, 85–110. New Brunswick, NJ: Rutgers University Press, 1985.

Parkinson, Gavin. "Emotional Fusion with the Animal Kingdom: Notes Towards a Natural History of Surrealism." *The Art of Evolution: Darwin, Darwinisms, and Visual Culture*, edited by Barbara Larson & Fae Brauer, 262–287. Hanover, NH: Dartmouth College Press, 2009.

Péladan, Joséphin "Sâr." "Manifesto and Rules of the *Salon de la Rose + Croix*." In *Art in Theory, 1815–1900: An Anthology of Changing Ideas*, edited by Charles Harrison, Paul Wood & Jason Gaiger. Translated Robert Pincus-Witten, 1054–1060. Oxford: Blackwell Publishing, 1998.

Pépin, François. "Diderot et la conjecture expérimentale, ou comment concilier l'extravagance de l'enthousiaste et la prudence de l'authentique genie." *MLN* 129, no. 4 (September 2014): 756–779.

Perlo, Katherine. "Marxism and the Underdog." *Society & Animals* 10, no. 3 (2002): 303–318.

Perry, Gill, Francis Frascina & Charles Harrison. *Primitivism, Cubism, Abstraction: The Early Twentieth Century*. New Haven, CT: Yale University Press, 1993.

Pippin, Robert M. *After the Beautiful: Hegel and the Philosophy of Pictorial Modernism*. Chicago: University of Chicago Press, 2014.

Pointon, Marcia. "The Representation of Time in Painting: A Study of William Dyce's *Pegwell Bay: A Recollection of October 5th, 1858*." *Art History* 1, no. 1 (March 1978): 99–103.

Prinzhorn, Hans. *Artistry of the Mentally Ill: A Contribution to the Psychology and Psychopathology of Configuration*. Translated by E. V. Brockdorff. New York: Springer, 1972.

Prodger, Phillip. *Darwin's Camera: Art and Photography in the Theory of Evolution*. New York: Oxford University Press, 2009.

Prodger, Phillip. "Ugly Disagreements: Darwin and Ruskin Discuss Sex and Beauty." In *The Art of Evolution: Darwin, Darwinisms, and Visual Culture*, edited by Barbara Larson & Fae Brauer, 40–58. Hanover, NH: Dartmouth College Press, 2009.

Przybyszewski, Stanislaw. "Psychic Naturalism (the Work of Edvard Munch)." In *Art in Theory, 1815–1900: An Anthology of Changing Ideas*, edited by Charles Harrison, Paul Wood & Jason Gaiger. Translated by Nicholas Walker, 1044–1059. Oxford: Blackwell Publishing, 2000.

Ranki, Kreeta. *Eläinten hyvinvointi uuden ajan alun eläinkäsitysten valossa*. Turku: Turun yliopisto, 2010.

Raphael, Max. *Prehistoric Cave Paintings*. Translated by Norbert Guterman. New York: Pantheon Books, 1945.

Ray, Nicholas 2014. "Interrogating the Human/Animal Relation in Freud's *Civilization and its Discontents*." *Humanimalia* 6, no.1 (2014): 10–40.

Redon, Odilon. "Suggestive Art." In *Art in Theory, 1815–1900: An Anthology of Changing Ideas*, edited by Charles Harrison, Paul Wood & Jason Gaiger. Translated by Akane Kawakami, 847–852. Oxford: Blackwell Publishing, 1998.

Reinach, Salomon. *Apollo: histoire générale des arts plastiques*. Paris: Librairie Hachette, 1904.

Reinach, Salomon. *Cults, Myths and Religions*. Translated by Elizabeth Frost. London: David Nutt, 1912.

Reinach, Salomon. "Paul Girop et Emile Massenat, les stations de l'âge du Renne dans les vallées de la Vézère et de la Corrèze." *Revue Archeologique*, 3ème série (1899): 476–478.

Richardson, Jonathan. *An Essay on the Theory of Painting*. London: A.C., 1725. Accessed May 15, 2014. https://ia802304.us.archive.org/35/items/essayontheoryofp00rich/essayontheoryofp00rich.pdf.

Rilke, Rainer Marie. *Auguste Rodin*. Translated by Victoria Charles. New York: Parkstone International, 2012.

Rilke, Rainer Maria. *Duino Elegies*. Translated by Stephen Cohn. Manchester: Carcanet, 1989.

Rilke, Rainer Maria. *Hiljainen taiteen sisin: kirjeitä vuosilta 1900–1926*. Helsinki: Ntamo, 2012.

Rilke, Rainer Maria. "In Celebration of Me (1909)." Translated by Annemarie S. Kidder. Accessed November 19, 2016. https://en.wikiquote.org/wiki/Rainer_Maria_Rilke.

Rilke, Rainer Maria. *Letters of Rainer Maria Rilke*. Translated by Jane Bannard Greene & M.D. Herter Norton. New York: W.W. Norton & Company, 1947. Accessed May 8, 2015. http://archive.org/stream/lettersofrainerm030825mbp/lettersofrainerm030825mbp_djvu.txt.

Rilke, Rainer Maria. "From *Letters on Cézanne*." In *Art in Theory, 1900–2000: An Anthology of Changing Ideas*, edited by Charles Harrison & Paul Wood. Translated by Joel Agee, 36–39. Oxford: Blackwell Publishing, 2003.

Rilke, Rainer Maria, *The Notebooks of Malte Laurids Brigge*. Translated by William Needham. Accessed November 19, 2016. https://archive.org/details/TheNotebooksOfMalteLaurids Brigge.

Rimbaud, Arthur. *Complete Works, Selected Letters*. Translated by Wallace Fowlie. Chicago: University of Chicago Press, 2005.

Ringbom, Sixten. *Pinta ja syvyys*. Translated by Rakel Kallio, Esko Kukkasniemi & Paula Nieminen. Helsinki: Kustannusosakeyhtiö Taide, 1989.

Rothko, Mark. "The Romantics Were Prompted." In *Writings on Art*, edited by Miguel López-Remiro, 58–59. New Haven, CT: Yale University Press, 2006.

Rousseau, Jean-Jacques. *"The Discourses" and Other Early Political Writings.* Edited by Victor Gourevitch. Cambridge: Cambridge University Press, 1997.

Rousseau, Jean-Jacques. *Émile, or On Education.* Translated by Barbara Foxley. Sioux Falls, SD: NuVision Publication, 2007.

Rousseau, Jean-Jacques. "Essay on the Origin of Languages." *On the Origin of Language.* Translated by John H. Moran & Alexander Gode, 1–83. Chicago: University of Chicago Press, 1986.

Rousseau, Jean-Jacques. *On the Origin of Inequality.* Translated by G.D.H. Cole. New York: Cosimo, 2005.

Rousseau, Jean-Jacques. *The Social Contract & Discourses.* Edited by G.D.H. Cole. London: J.M. Dent, 1920.

Rousseau, Pascal. "Confusion des sens: Le débat évolutionniste sur la synésthesie dans le débuts de l'abstraction en France." *Les Cahiers du Musée national d'art moderne* 74 (hiver 2000–2001): 4–33.

Sartre, Jean-Paul. *Being and Nothingness: The Principal Text of Modern Existentialism.* Translated by Hazel E. Barnes. New York: Washington Square Press, 1992.

Sartre, Jean-Paul. *"What is Literature?" and Other Essays.* Cambridge, MA: Harvard University Press, 1988.

Sawyer, Paul L. "Ruskin and Tyndall: The Poetry of Matter and the Poetry of Spirit." In *Victorian Science and Victorian Values: Literary Perspectives*, edited by James Paradis & Thomas Postlewait, 217–246. New Brunswick, NJ: Rutgers University Press, 1985.

Schaeffer, Jean-Marie. *Art of the Modern Age: Philosophy of Art from Kant to Heidegger.* Translated by Stephen Rendall. Princeton, NJ: Princeton University Press, 2010.

Schaeffer, Jean-Marie. *La fin de l'exception humaine.* Paris: Gallimard, 2007.

Schelling, Friedrich. "Concerning the Relation of the Plastic Arts to Nature." In *Art in Theory, 1648–1815: An Anthology of Changing Ideas*, edited by Charles Harrison, Paul Wood & Jason Gaiger. Translated by Michael Bullock, 934–941. Oxford: Blackwell Publishing, 2000.

Schlegel, Friedrich. *Friedrich Schlegel's Lucinde and the Fragments.* Translated by Peter Wirchow. Minneapolis: University of Minnesota Press, 1971.

Simms, Matthew T. "The Goncourts, Gustave Planche, and Antoine-Louis Barye's *Un Jaguar dévorant un lièvre.*" *Nineteenth-Century French Studies* 38, nos. 1 & 2 (Fall–Winter 2009): 67–84.

"Sketching the Universe: Artists' Responses to D'Arcy Thompson." University of Dundee. Accessed November 17, 2016. www.dundee.ac.uk/museum/exhibitions/zoology/sketching/

Smith, Barry & David Woodruff Smith, ed. *The Cambridge Companion to Husserl.* Cambridge: Cambridge University Press, 1995.

Spies, Werner. *Max Ernst—Loplop: The Artist in the Third Person.* New York: George Braziller, 1983.

Spinoza, Benedict de. *Complete Works.* Translated by Samuel Shirley. Cambridge: Hackett, 2002.

Steiner, Gary. *Anthropocentrism and Its Discontents: The Moral Status of Animals in the History of Western Philosophy.* Pittsburgh, PA: University of Pittsburgh Press, 2005.

Steiner, Rudolph 2010. *Haeckel, The Riddle of the Universe, and Theosophy.* Whitefish, MT: Kessinger Publishing.

Stubbs, George, Thomas Villiers, Earl of Clarendon & George Frederick Laurence Hyde Villiers, Earl of Clarendon. *No. I–III of a Comparative Anatomical Exposition of the Structure of the Human Body, with That of a Tiger and Common Fowl: In Thirty Tables, Two Copper-Plates to Each Table, 20 Inches by 15.* London: J. White, 1803–1806.

Tapié, Michel. "From *An Other Art.*" In *Art in Theory, 1900–2000: An Anthology of Changing Ideas*, edited by Charles Harrison & Paul Wood, 629–631. Oxford: Blackwell Publishing, 2003.

Tisdall, Caroline. *Coyote*. Munich: Schirmer/Mosel, 2008.

Tolstoy, Leo. *What is Art? What is Religion?* Translated by Aylmer Maude, V. Tchertkoff & A.C. Fifield. Milford, PA: Wildside Press, 2008.

Tomkins, Calvin. *Bride and the Bachelors*. London: Penguin, 1968.

Tomkins, Calvin. *Duchamp: A Biography*. New York: The Museum of Modern Art, 2014.

Uexküll, Jakob von. *A Foray into the Worlds of Animals and Humans: With a Theory of Meaning*. Translated by Joseph D. O'Neill. Minneapolis: University of Minnesota Press, 2010.

Waddington, C.H. *Behind Appearance: A Study of the Relations between Painting and the Natural Sciences in This Century*. Cambridge, MA: MIT Press, 1970.

Wellek, René. *A History of Modern Criticism: The Later Eighteenth Century*. Cambridge: CUP Archive, 1981.

Wild, Markus. *Die anthropologische Differenz. Der Geist der Tiere in der frühen Neuzeit bei Montaigne, Descartes und Hume*. Berlin: Walter de Gruyter, 2006.

Winckelmann, Johann Joachim. *Reflections on the Imitation of Greek Works in Painting and Sculpture*. London: A. Millan, 1765. Accessed May 15, 2014. https://archive.org/details/reflectionsonpai00winc.

Wolloch, Nathaniel. *Subjugated Animals: Animals and Anthropocentrism in Early Modern European Culture*. New York: Humanity Books, 2006.

Worringer, Wilhelm. *Abstraction and Empathy: A Contribution to the Psychology of Style*. Translated by Michael Bullock. Chicago: Ivan R. Dee, 1997.

Zgórniak, Marek. "Frémiet's *Gorillas*: Why Do They Carry Off Women?" *Artibus et Historiae* 27, no. 54 (2006): 219–237.

Zola, Émile. "The Experimental Novel." In *Documents of Modern Literary Realism*, edited by George J. Becker, 161–197. Princeton, NJ: Princeton University Press, 2015.

Index

9781032339535